Praise for Christopher McDougall's

Running with Sherman

"A charming story of rehabilitation, teamwork, athleticism, and the healing power of love between people and their four-legged companions. . . . *Running with Sherman* [is] a fun and inspiring read, not just for runners, but for anyone who believes in the healing power of the human-animal bond."

—*New York Journal of Books*

"Both hilarious at times and touching . . . written with pizzazz."

—*Inc.*

"A smart critique of the culture of conventional American sports."

—*Outside*

"I've always believed that people who didn't love miniature donkeys are going to hell. I think people who don't love this book may be, too. This is more than just a lovely tale about a wounded animal. It is the kind of story that breaks your heart and mends it all in the space of a paragraph or even a line. Besides, it's about a MINIATURE DONKEY."

—Rick Bragg, bestselling author of *The Best Cook in the World*

"Christopher McDougall is back with another inspirational of borderline insane people racing inhumanly long distances. . . . This time Chris isn't just pounding the pavement with his fellow homo sapiens. In *Running with Sherman*, he recounts his efforts to rehabilitate the body, mind, and spirit of a weary donkey by entering them both in the pack burro racing World Championship."

—Literary Hub

"A charming story about animals and people in Amish Country."
—Temple Grandin, author of *Animals Make Us Human*

"What an insane, amazing story. I genuinely cried so many times throughout it and cried the happiest tears at the end. *Running with Sherman* is an inspiring love letter to the beautifully broken and the immeasurable power of healing together. It is a meaningful reminder that no one is ever beyond repair. This book motivated me to test my limits and made me feel deeply ashamed my work ethic and athletic prowess are only half that of a farm animal. Sherman the Donkey is my personal superhero."
—AJ Mendez Brooks, aka AJ Lee,
WWE superstar and author of *Crazy Is My Superpower*

"Sherman's transformation from dying donkey to confident runner involved a circle of family, friends, neighbors, and a few feisty donkeys, each of whom McDougall portrays in affectionate, vivid detail. . . . A charming tale of a resilient donkey and a community's love."
—*Kirkus Reviews*

"McDougall is a charming, enthusiastic storyteller. . . . Runners and animal lovers alike can find inspiration in this story of the ways in which humans and animals connect."
—*Publishers Weekly*

"You can't help but fall in love with Sherman the determined donkey. *Running with Sherman* is charming and hilarious, heartfelt and sincere. It's a wonderful read."
—*The Maine Edge*

Christopher McDougall

Running with Sherman

Christopher McDougall covered wars in Rwanda and Angola as a foreign correspondent for the Associated Press before writing his bestselling book, *Born to Run*. His fascination with the limits of human potential led him to create the *Outside* magazine web series *Art of the Hero*. He currently lives with his wife, two daughters, and a farmyard menagerie in Lancaster County, Pennsylvania.

www.chrismcdougall.com

ALSO BY CHRISTOPHER McDOUGALL

Born to Run

Natural Born Heroes

Running with Sherman

Running with Sherman

How a Rescue Donkey
Inspired a Rag-tag Gang of Runners
to Enter the Craziest Race in America

Christopher McDougall

VINTAGE BOOKS
A DIVISION OF PENGUIN RANDOM HOUSE LLC
NEW YORK

FIRST VINTAGE BOOKS EDITION, JULY 2020

The Library of Congress has cataloged the Knopf edition as follows:
Name: McDougall, Christopher, author.
Title: Running with Sherman : the donkey with the heart of a hero / Christopher McDougall.
Description: First edition. | New York : Alfred A. Knopf, 2019.
Identifiers: LCCN 2019009852
Subjects: LCSH: Sherman (Donkey). | Human-animal relationships—United States. | Pack burro racing—United States. | Donkeys—Training—United States.
Classification: LCC SF361 .M37 2019 | DDC 636.1/82—dc23
LC record available at https://lccn.loc.gov/2019009852

Vintage Books Trade Paperback ISBN: 978-0-525-43325-5
eBook ISBN: 978-1-5247-3237-0

Author photograph © Matt Roth
Book design by Maggie Hinders

www.vintagebooks.com

Printed in the United States of America
10 9 8

On Sherman's behalf,

this book is dedicated to the three women

who brought the joy and adventure into our lives:

Mika, Maya, and Sophie.

To achieve great things, two things are needed:

a plan, and not quite enough time.

—LEONARD BERNSTEIN

Running with Sherman

1

Shadow in the Dark

I knew something was wrong the second the pickup truck pulled into our driveway. I'd been waiting for Wes for more than an hour, and now, before he even came to a stop, the look in his eye warned me to brace myself.

"He's in rough shape," Wes said as he got out of the truck. "Rougher than I thought." I've known Wes for more than ten years, nearly from the day my wife and I first uprooted ourselves from Philadelphia to live on this small farm in Pennsylvania Amish country, and I'd never seen him so grim before. Together, we walked behind the pickup and pulled open the trailer doors.

I took a look inside, then immediately grabbed in my pocket for my phone. Luckily, I had the number I needed.

"Scott, you've got to get over here. This is really bad."

"Okay," Scott said. "You just make him comfortable and I'll be over in the morning."

"Yeah. No. I think you'd better, um, have to—" I paused a sec to untangle my tongue. Scott was the expert, not me, but I didn't think we had many mornings left to work with. I tried again to tell him what I was looking at.

Inside the trailer was a gray donkey. Its fur was crusted with dung, turning his white belly black. In places the fur had torn away, revealing raw skin almost certainly infested with parasites. He was barrel-shaped and bloated from poor feed and his mouth was a mess, with one tooth so rotten it fell right out when touched. Worst of all were its hooves, so monstrously overgrown they looked like a witch's claws.

"Scott, seriously. You've got to see this."

"Don't worry," Scott said. "I've seen it all. Catch you in the morning."

The donkey belonged to a member of Wes's church. Wes is a truly wonderful person to begin with, and as a Mennonite, he's committed by faith to helping anyone in need—or, in this case, any creature. Wes had discovered that one of his fellow churchgoers was an animal hoarder who kept goats and a donkey penned in squalor in a crumbling barn. The hoarder was out of work, so his family was suffering from his fixation as well; money needed for food and rent was going for animal feed instead. Wes and several church elders had tried to persuade the hoarder to relinquish his pets, but he wouldn't budge. Finally, Wes took a deep breath and bent his iron-hard honesty to the limit. What if, he asked the hoarder, we take the animals away for two years? Just two years. We'll give them to a good family and get them healthy, and that will give you time to put up some fences and clean out those stalls. It wasn't really a *lie*, Wes told himself. More like a hope—the hope that two years would be long enough for the hoarder to forget these poor animals and get on with his life.

"Give it a try?" Wes persisted.

"Okay," the hoarder replied. "But they have to go to a good family."

Wes got on the job at once. The goats were easy to place—someone in Lancaster can always use a free lawn mower—but donkeys are tough. They're famously ornery, known for biting and kicking, and serve no purpose on a working farm. They can't be milked or butchered or, in many cases, even ridden. Keeping them in hay and feed can be expensive, and that's before you're shelling out for dental care and deworming and vaccinations.

So why did I want him?

I didn't. Not when I got a good look at him, that was for sure. As transplanted city folk who knew zip about farm life when we moved to the country, my wife and I had gotten a kick out of trying our hand with a few starter animals. First up was a stray black cat that appeared at the back door, and when it survived and stuck around we advanced to some backyard chickens, and then a foster sheep that we took on loan from an Amish neighbor to see if we could handle it, like a kindergartner bringing home the class's pet turtle for the weekend. Wes owns the farm next to ours, and when he told me about the donkey he was trying to rescue, I figured why not? We could just turn it loose out back and let the kids feed it apple cores. I wasn't making any promises till we saw it, though, which was fine by Wes; the donkey's owner, he said, was a bit of a handful who felt the same way about me.

So one afternoon, my two young daughters and I headed over to the hoarder's house to check things out. Secretly, that was just our cover story; the girls and I had already made up our minds before we'd even gotten in the car that unless this thing was a rampaging maniac, we were bringing it home. During the drive, we schemed up ways to talk Mommy into this operation and debated names for our future pet.

"Skullcrusher?"

"No!"

"Zorro?"

"NO! Actually, maybe."

But our happy chatter died once we arrived. The hoarder's barn was sagging in a field of mud, looking like a sneeze would bring it down. We slogged inside, straining to see in the gloom and to pull our boots out of the sucking muck. A hard rain had fallen the day before, flooding one of the pens so badly that two of the goats had to stand on straw bales to stay above water. Next to the goats was another stall, this one as dark and tiny as a dungeon cell.

Inside, another creature was barely visible against the back wall. The hoarder called and whistled, holding out a handful of feed.

Slowly, a shadow detached itself from the darkness. Its long ears rose, twitching nervously, as it struggled to take a step toward us. The donkey was mired nearly to its knees in manure and rotten straw, and so cramped by the narrow stall that it could barely turn around. The hoarder poured the feed into my daughter's hand. She held it out, and the donkey stretched toward us to gently snuffle it from her palm. My daughters and I stared at him in silence. We didn't care anymore about getting a pet. All we cared about was getting him out of there.

The hoarder agreed to let us have him. But overnight, he changed his mind. When Wes showed up the following morning with the trailer, the hoarder dug in his heels and said no, the donkey was family. Family had to stay together.

"Remember, it's just till he gets better. Just for two years," Wes repeated, over and over, until finally the hoarder relented and opened the stall door. That's when Wes discovered the donkey's hooves had been so badly neglected that he could barely walk. Together, Wes and the hoarder struggled, one step at a time, to ease the sick animal out of the dark barn, into the daylight, and off to his new home.

"How do we get it off the trailer if it can't walk?" I asked Wes, dreading the possibility that he might actually have an answer. I held my breath, silently urging him to say it couldn't be done and he'd have to speed the donkey off to some kind of sanctuary or animal urgent-care or wherever it is they handle the hopeless cases.

"Slowly, I guess," Wes replied. He took hold of the threadbare green halter around the donkey's head and very gently pulled forward. What should I do, get behind and push? That seemed aggressive. And besides, from what little I knew about donkeys, trying to shove it could put me right in the danger zone for an angry hoof to the kneecap. Maybe I could kind of *hoist* it a little?

Sherman arrives.

I threw my arms around the donkey's back and cradled its belly with both hands, trying awkwardly to lift the weight off its ailing hooves. I was ready to jump if it kicked, but it didn't seem to have any fight at all. It looked dazed, more like a moldy toy hauled out of a basement than a living creature. Gingerly, it took one slow step after another. It walked when we urged it and stopped when we didn't, as if it no longer remembered a time when it could think—and move—for itself. When we got to the end of the ramp, the donkey didn't even start munching the tasty green grass; it froze back into a stuffed animal, head hanging and motionless.

Wes had to scoot. He had 150 dairy cows waiting to be milked back home, and his last bout of hostage negotiations with the hoarder had put him way behind schedule. Wes wished me well and promised to stop by the next day to see how we and the patient were making out. I put a bucket of fresh water and some hay in front of the donkey, which still hadn't budged, and checked my watch. My daughters would be getting home from school soon. I wanted to greet them with an action plan, something that would soften the shock of the donkey's appearance and let them know the animal would eventually be all right, but I didn't have a clue. We'd wanted to help a creature in need, but this kind of creature—and this kind of need—was way beyond anything I'd imagined.

2

Hacksaw Surgery

Early the next morning, our savior rolled into the driveway. "Don't worry," I'd reassured the girls the night before. "Scott will know what to do." Sure enough, Scott hopped out of his truck with a confident grin—which quickly faded.

"I've seen it all," he said. "But not this."

By day, Scott is a sales rep for Dansko, the Pennsylvania company that makes those clogs that chefs and dancers love. In the evenings, he shifts his focus from feet to hooves, his real passion. Scott grew up in upstate New York and paid his way through college by learning to fit horses for shoes. After he moved to our neighborhood in Lancaster County—home of America's largest Amish community—he became the go-to guy whenever local farmers needed help with their big work mules and buggy horses.

Some weekends, Scott and his wife, Tanya, would wander around horse auctions as unofficial animal advocates, speaking up when they spotted a horse that needed care. Once, Tanya jumped in front of an auction trailer headed to the slaughterhouse, pulled out her wallet, and told the driver to name his price for a mini donkey she'd spotted in the back. The little donk was so far gone that the driver gave it to her for free. Tanya thought she could heal it, and she was right. Soon, tiny Matilda was trotting along when Tanya and Scott took their carriage horses out for a drive. But this creature that turned up in my driveway was even worse off than Matilda had been.

"Man, how did this happen?" Scott asked.

"Hoarder," I replied.

"Jeez, this is . . ." Scott began. He stopped to think for a sec. "Look, the most humane thing might be to put him down now."

The hooves, he explained, were a death sentence. Donkeys usually keep their hooves naturally pumiced by foraging for long miles over rocky ground. But if you pen them up on soggy straw, or even leave them standing around all the time in a grassy meadow, their hooves will eventually curl like the nails of a Hindu holy man. Once they're deformed, the damage can be irreversible and lead to an excruciating death: Because equines have unusually small stomachs, most digestion takes place when their intestines are churned by the rocking motion of walking. Hobble them, and it's only a matter of time before waste matter blocks their guts until the animal is torn apart from the inside.

"That's a horrible way to die," Scott said. "Unless . . ." He paused to think for a moment. "Do you have a hacksaw?"

I ran to the shed and fetched one. Scott tied the donkey's halter to a fence post. "Hi, buddy," he said, stroking the donkey's ears. "Ever seen one of these before?" He held the hacksaw under the donkey's nose so he could sniff it. "Now here's what we're going to do," Scott said. He began explaining to the donkey the details of the procedure he was about to perform. It made me cringe, but the donkey's ears swiveled toward Scott as if it was listening intently.

"I want him to get used to my voice before we start," Scott told me. "Donkeys are very self-protective, way more than horses. They don't like surprises." And what Scott had in mind was going to be rough: as last-ditch emergency surgery, we were going to hold each hoof, one after the other, and saw through it like a tree limb. If Scott could cut off at least half of each hoof, he could then try shaping the hoof with his steel clippers and a rough file. Imagine going to the dentist with four cavities and finding out that each tooth had to be drilled not once but three times in a row, except you've never seen a dentist in your life and for all you know, the guy with the drill is some psycho grabbing you by the jaw. That's what the donkey and I were in for.

"Ready?" Scott asked.

"Me? Or him?"

"Both of you. Grab that leg and hold on tight."

We got to work. I leaned against the donkey's flank, pressing him securely between my body and the fence, while Scott straddled the donkey's leg and secured the first hoof between his knees. Slowly and carefully, he notched into it with the saw. Once he got a groove going he really leaned in, pushing and pulling through a hoof that was as tough as a car tire. Sweat began pouring down his face, but even though he was panting, Scott kept talking to the donkey in a calm, affectionate voice.

"Doing good, buddy?" he said. "We're almost through number one." Every muscle in the donkey's body was so tense it seemed he was ready to erupt, but amazingly, he stood firm. Scott finally put down the saw and straightened up, wiping his soaking forehead.

"What do you think of this?" he asked. He held up a severed hunk of hoof that was nearly the size of my foot. The smell was repulsive, as if the hoof had already begun to rot while attached to the donkey's foot.

"I can't believe he let you do that," I said. "He must still be shell-shocked from the move."

"Maybe. But he's a really good guy," Scott said, ruffling the fur on the donkey's head. "See how his ears shift as we're talking?" Sure enough, the donkey's ears rotated from me to Scott every time we opened our mouths. Sometimes he'd even split them in opposite directions, pointing one ear at me and the other at Scott like a cop directing traffic.

"He's paying serious attention, and he's decided we're on his side," Scott said, as he knelt to begin sawing the other front foot. "But don't let up. It's only going to get rougher now."

We had to tackle the hind hooves next, and Scott told me that

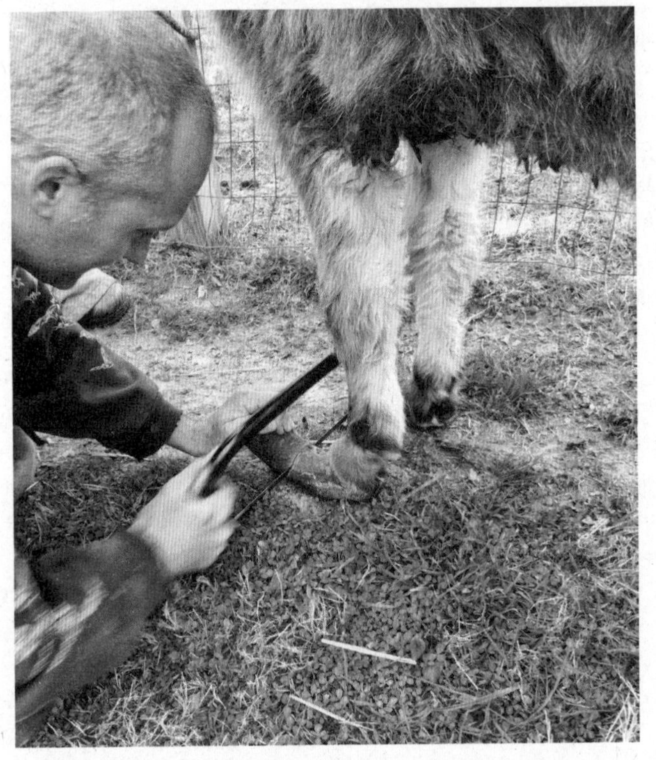

Scott working on Sherman's hooves

no donkey in the world likes someone getting behind him. "That's their biggest primal fear," Scott said. Out in the wild, donkeys are pretty tough to kill. They're herd animals and stick super-tight to the group, so any predator hoping for a donkey dinner has to weigh its chances of coming out alive against a mob of kicking, biting, 700-pound beasts who've been known to stomp lions to death. But donkeys are still vulnerable to sneak attacks; any straggler who lingers a bit to graze can suddenly find wild dogs leaping on its back and taking out its throat. This little gray donkey was weak and sick, but deep inside its DNA was a 10,000-year-old survival instinct as sharp as any Army Ranger's, always alert to cover its six and kick for its life against anything it couldn't see.

Scott picked up the saw. Softly, he laid his free hand on the donkey's back leg. "Good boy—" he began, then jerked back as the donkey's leg shot out.

"That's what I mean," he said, bending down for the saw he'd dropped. "He can break your leg before you see him move." His own little donkey, Matilda, once dealt with a snapping dog by blasting it so hard that the dog's leg had to be amputated.

"Hold him tight and let's try this again," Scott said. I pressed my chest against the donkey's ribs, pushing him as firmly as I could against the fence. Scott rubbed the donkey's head reassuringly, then worked his way down its body, scratching and massaging its back bit by bit until he reached the haunches. He smoothed his hands down the back leg, easing his way toward the hoof. The donkey looked like a robbery victim, its ears sticking straight up like it was being held at gunpoint, but it remained frozen while Scott slowly picked up a rear hoof. Whether we pulled this off or not, I couldn't have been more impressed by Scott's bedside manner. Even though he was sweating like a blacksmith and expecting to get his ribs cracked at any moment, he continued crooning to the donkey as if he were offering a juicy apple instead of wielding a hacksaw.

Finally, the last big chunk of hoof dropped off. Scott took a breather and wiped the grime off his face, but the ordeal wasn't over. From his side pouch, Scott pulled a freakishly huge pair of steel clippers that looked like they'd been designed by Leatherface for use in his murder van. Scott warmed up with a few practice snips in the air, then leaned back in toward Sherman and began nipping away with expert precision, doing his best to shape the sawn stubs into something resembling a healthy animal's hooves. After he'd pared each hoof, he pulled out a foot-long metal file for a final smoothing.

"Done!" Scott announced. He flopped back on the grass in exhaustion, sucking in deep breaths of relief. His T-shirt and jeans, spotless when he'd arrived, now looked like they'd been dug out of a swamp. He'd barely gotten comfortable when he shot back up, alarmed by what Sherman was doing:

Nothing.

"Not great," Scott said. "Not great." We'd just put this animal through the equivalent of two hours of dental surgery, and instead of making a break for daylight, the little gray donkey was standing exactly where we'd left him. Now that his hooves were trimmed and he was free to go, why wasn't he hightailing it out of our reach?

The two of us watched the donkey, mentally urging him to get a move on, but after a good long while, he hadn't taken a single step. "I don't know," Scott said, his voice sounding weary and resigned. "If he's not walking by tomorrow, all we can do is make him comfortable before he goes."

Comfort was his wife's department, and it wasn't long before Tanya was roaring up our driveway in her dusty old SUV. She charged into action with her medical kit and shears, swiveling her head back and forth as she alternately crooned to the donkey and barked commands back at me.

"Good donkey!" she purred. "Good—" She paused. "What's his name?"

"Um . . ." I knew the stakes were high and I didn't want to blow it. We'd messed up in this department before and I was still paying the price. Two of the first goats we ever got were named for words my daughter had recently come across in books: "Bamboozle" and "Skeedaddle." Even though Bamboozle and Skeedaddle had four acres of lush grass and tasty weeds to keep them munching, they became master escape artists that spent most of the day prowling the fence like twin El Chapos searching for places to tunnel to freedom. After a few months of successful breakouts, they didn't even bother squirming under the wire anymore; they just took to the air, sailing over the five-foot fence and ignoring the two hundred acres of neighboring cornfield all around them to head, naturally, straight for the road, where they could wander out in front of the school bus and give me heart attacks.

Finally, I surrendered and sold Skeedaddle and her sister, Lulu, to a farmer for his grandkids to raise and gave Bamboozle to our nearest Amish neighbors. The next morning, we looked out the window and found Bamboozle looking back at us. He'd slipped away from his new owners and trotted half a mile down the road to escape back *into* our yard. The girls were delighted—they adored Bamboozle and wanted to keep him—but I'd hit my limit for lunging and sprawling after him like a circus clown. Luckily, the Amish kids must have figured out a way to stop him, because after I brought Bamboozle back that time, he never escaped again. I stopped by a few days later to ask how they'd finally outsmarted him, and they looked confused.

"Oh, you mean *Fred*," one of them said. "That's what we call him now."

Yes, Fred—the same name as two-thirds of the elderly uncles currently sleeping off lunch in their Barcaloungers. I didn't know how giving Bamboozle a chill new name turned him into a chill new goat, but you'd think I would have given it a try for myself.

Nope. Instead, we went on to name a stray kitten "Smartycat," and watched this strictly outdoor cat become a genius at darting into the house and vanishing into sock drawers just when we were trying to leave on a family trip. Smartycat lived a good, long life with us, and when she died, she was replaced by "Evil Eye," another stray that basically forced that name on herself because of her creepy Satanic serpent's stare. Evil Eye was (and remains) so mean that our other cats, all half-wild scrappers themselves, are afraid to approach any of the three food bowls until Evil Eye is fed, full, and out of sight.

So when it came to naming this sick donkey, I wasn't going to mess around. He was already in a fight for his life, and he didn't need me making things worse by giving him a name with even the hint of a hex. The girls had made a suggestion the night before, and after examining it from every angle, I couldn't spot any danger. I decided to run it by Tanya.

"We're thinking of calling him, um . . . Sherman?" I said. We'd recently seen the movie *Saving Mr. Banks*, and we'd gotten a kick out of the happy-go-lucky, songwriting Sherman brothers. Who doesn't love "Let's Go Fly a Kite"?

Tanya couldn't care less about Disney films or voodoo hexes. She was in full emergency room mode, and for her, a name was just another surgical tool. "Good Sherman!" she crooned, clicking on her big-toothed shears. She got to work on the donkey's stinking, matted fur, peeling it away in strips. Every once in a while, she called back over her shoulder for some item she needed from the house: Rags! Baby shampoo! Get the hose!

"As soon as I'm finished, you'll need to soak him down and lather him from nose to tail," she commanded. "Give him a good shampooing. He's not going to like it at first, but you need to stick it out. You keep cleaning him until you get all this filth off his skin." Suddenly, Tanya clicked off her shears and turned to face me.

"Look," she said. "If he makes it, you can't just stick a ribbon on his tail and leave him standing in a field like Eeyore. He's been abused and abandoned, and that can make an animal sick with despair. You need to give this animal a purpose. You need to find him a job."

A *job*? What was I going to do with a donkey, prospect for gold? Pioneer westward? But before I even asked what she meant, I got an idea. Nah, that's ridiculous, I thought to myself, and kept my mouth shut. No way was I going to share this with Tanya and look even more helpless and out of my depth than I already did. Still, the more she worked on the grim wreckage of Sherman's body, the more I circled back to this fantasy. I couldn't let it go, and I realized why: focusing on a glorious fairy tale was a lot more pleasant than the ugly reality that was kicking us in the face.

And that's when it dawned on me that my incompetence did have an upside: it cut both ways. Since I had no idea how sick Sherman was, *I didn't know how strong he was either*. For all I knew, there might just be a fighter in there, a fierce warrior spirit buried deep inside that was laying low until it gathered the strength to start surging through Sherman's veins. And if Sherman found his way back to life, maybe I had something for him that was even better than a job: a wild adventure that the two of us could tackle together, side by side.

But first we had to keep him alive.

3

No One Likes Us, We Don't Care

"Oh, crap!" Tanya suddenly realized it was nearly three p.m. "Late for school."

She grabbed her shears and gear and moments later her SUV was spitting gravel as Hurricane Tanya disappeared down the driveway. Every morning and afternoon, Tanya was the driver for local Amish children who lived too far from their one-room schoolhouse to make the trip on foot. After delivering them home, she had a full evening of chores with her own animals, which included three donkeys, two carriage horses, one goat, one pig, a wading pool full of ducklings, and a rescue horse she'd saved from slaughter so she could teach a teenage neighbor how to ride. She wouldn't be able to check on Sherman again till morning.

"What do we do now?" asked my wife, Mika. We stood at the fence, waiting to see if Sherman would move.

Nothing.

"Either he's getting better, or—" I glanced around to make sure the kids couldn't hear. "Or we're on death watch. Tanya said at this point, it's out of our hands."

Out of our hands. It gave me a sickening feeling, saying those words, because for one of the few times in my life, it was true. There was no one else to call, no further treatment to try, no friend to seek for advice. That little spark of hope I'd felt a minute ago faded away, replaced by the chest-squeezing grip of doom you get when your car spins on ice. Sherman was alone inside this tunnel, and he was either going to walk out the other end on his own or disappear into the darkness.

I just wished I knew what was going through his mind. If there was no way to pull him back, at least we could ease his exit with kindness and care. But how could we bring him peace when we had no clue what he was thinking? Was he fighting for his life, or giving up? Did he see me as his friend, or as just another tormentor? The first rule of healing is "Do no harm," but Sherman was making me realize that I knew so little about animals, I couldn't tell if I was soothing or scaring him.

Mika and I weren't just surprised to be in this predicament. We still couldn't believe we were in this zip code.

I grew up just outside of Philadelphia, in the working-class suburbs where the El tracks and row houses of West Philly gave way to the big families and small backyards of Upper Darby. My only contact with country livin' came from books; I was so obsessed with *My Side of the Mountain* that I ran away from home at age nine with only a Wham-O boomerang, fully intending to live in a hollow tree in the woods and hunt with a hawk like Sam Gribley did. Around one o'clock in the morning, the state police found me six miles from home in a patch of woods near Springfield Mall and hauled me back for a parental smackdown that was epic enough to put an end to any future walkabouts.

After that night, I was rarely far from the company of at least 1.5 million neighbors. I went to high school in North Philly and became a street-court rat, spending all my time outside of class roaming the city with my friends in search of pickup games. After college, I bounced around between jobs and cities before taking a leap overseas to see what life was like in Madrid. I taught English for a while and learned enough Spanish to finagle my way into an interview for a news reporting job with the Associated Press. I had no credentials for the job, but the bureau chief in Madrid, Susan Linnee, was a battle-hardened newswoman who scorned the hothouse-flower desk editors that New York headquarters kept sending her and preferred her own method for discovering street-savvy "talent in the rough," as she put it.

"The guy before you, what sold me was he looked like the lead singer of the Fine Young Cannibals," Susan told me. Luckily, the Cannibal turned out to be such a natural that within a year, he was recruited to become a war correspondent in Bosnia. Someone had to replace him, pronto, which was the only reason I got through the door for an interview. Susan grilled me for about an hour, and when my utter lack of experience became embarrassingly obvious, she abruptly stood up and called an end to the interview.

"I've heard enough," she said, sticking out her hand.

"Okay," I agreed, more than ready to beat it. "If you change—"

"We'll train you here for a week," she continued, already steaming ahead with her plans. "Then we really need you there."

"There?"

True, she had mentioned that the Cannibal was her Lisbon correspondent, but I naturally assumed they would transfer someone from Madrid and keep me at base to learn the ropes. I'd never been to Portugal in my life and didn't know a word of the language, but that wasn't my biggest problem. Civil war had just re-ignited in Angola, which didn't seem like any of my business until my new boss explained that as a former Portuguese colony, Angola literally became my business at the moment I shook her hand.

One month later, I was behind rebel lines in southern Africa, doing my best to stay alive while pretending I had any idea what I was doing. I was teamed with Guilherme, a Portuguese photographer who also spoke Spanish, so most of the time the only way I could gather information from Angolan soldiers was by way of a spoken-word Rube Goldberg machine: I'd feed my questions in Spanish to Guilherme, who would translate them into Portuguese for the soldiers, and then translate their answers back to me in Spanish so I could jot them down in English. Guilherme had his own work to do and really didn't have time for this nonsense, so he would listen to a soldier's long, teary-eyed saga and boil it down to "They shot lots of the bad guys."

"That's it?"

"Big picture."

Fine by me; the tighter the quote, the quicker I finished. Every day, I had to scout around, interviewing refugees, aid workers, and frontline fighters, then condense their info into AP news stories that needed to be sent to New York before sunset. Darkness was my deadline, because the only way to transmit from the field was with a satellite telex the size of a wheelie suitcase. You didn't want to be up on a hill looking for a signal with that thing at night; for a roving rebel soldier with an itchy trigger finger, the only thing visible against the dark sky would be the blinking green "SHOOT ME!" lights on my console. The second I hit Send, I slammed the cover shut and scuttled for safety.

Like the Cannibal, I managed to stick around long enough to get the hang of it. When massacres erupted in Rwanda two years later, I was assigned to embed with the Tutsi rebel army that was racing across the border to rescue civilians from the murderous militias. We were only a small band of reporters traveling with the Tutsis, and we got smaller by the day. One American correspondent was airlifted out when her photographer was shot through the legs and she had to stop the bleeding with her bare hands. A French radio journalist was stricken by cerebral malaria and barely survived. My photographer left after we entered a schoolhouse and found the bodies of dozens of young children who'd been hacked to death with machetes; the next morning, he found his hands were still trembling. When the Tutsis finally chased the murderers into Congo and the fighting died down, I was desperate for a rest. Instead, I couldn't sleep.

It was time to go home.

. . .

Maybe it was a bad idea to leave Lisbon, ditching a dream job in a beautiful seaside city, but it turned out I wasn't the only one making that mistake. I returned to Philly and quit the AP to scratch out a living as a freelance magazine writer. One afternoon, I was out for a run with Jen, a friend from the AP's Philly bureau, and she told me about a reporter from Hawaii who'd rotated in for a one-year stint. Island gal wasn't loving her new home, and Jen didn't have to tell me why: Philly can be cold and bitter, and that's just the people. If you're familiar with our monument to Frank Rizzo, one of the most brutal police chiefs in city history, or the time Santa Claus showed up at an Eagles game and we drilled his jolly old ass with snowballs, or the way Eagles fans and the Eagles themselves sang "We're from Philly, f***ing Philly, no one likes us, we don't care" after Philadelphia won the Super Bowl in 2018, you know it's not the warmest and fuzziest landing spot for strangers. It couldn't be easy for a homesick Hawaiian, so when Jen told me she was taking African dance classes, I thought I could cheer her up a little with some CDs I'd brought back from Angola.

Jen invited me to a dinner party that weekend. When I arrived, CDs in hand, I scanned the living room, searching for the brooding, heavyset Pacific Islander. I was still looking when a breathtaking woman with a warm, welcoming smile approached, looking like she'd just surfaced off a Tahitian island with a handful of pearls. I could barely stutter a greeting because my synapses were jammed by two colliding thoughts:

1. You *genius*, bringing those CDs.
2. Never, ever mention you thought all Hawaiians looked like NFL linemen.*

* I can't excuse my cultural ignorance except with the weak argument that a Hawaiian friend thinks the same way. When he first visited New York and saw the Statue of Liberty, he was disappointed: "I know Samoans who are bigger."

She told me her name was Mika,* and that's where our conversation ended. I handed over the music and then got as far away from her as I could, spending the rest of the evening in the corner looking at prints with my photography buddy, M'poze. I'd ruined enough first impressions in my time to grasp that when I surprised Mika with that gift, I'd hit my peak. This woman was so far out of my league, anything I said after that would just begin the process of scaring her off. When Mika brought me a plate of food a little later, I gave her a quick thanks over my shoulder and pivoted right back to M'poze's photo book. I stayed riveted to that thing for so long, even M'poze was getting bored. But if he was cornered, so was I. I was betting the house on the Tao of Steve.

I'd recently seen an indie film that offered the theory that the best way to attract someone is to follow the leads of both Zen Buddhism and those twin pillars of sexy self-command, Steve McQueen and the Six Million Dollar Man, Steve Austin. The Tao of Steve wasn't pickup-artist stuff; it was more like a guide to better living through impulse control, based on the premise that you get what you want only when you stop wanting it. Whenever you meet someone who makes your pulse race, you're supposed to follow three steps:

- Be desire-less
- Be excellent
- And be gone

Purely by dumb luck, I'd nailed the first two. I'd arrived looking like a hero, and if I didn't want to blow it, I had to keep my mouth shut and get out the door while Mika still thought I was kind and chill. So chill, in fact, that when the party was winding down and

* If you want to see for yourself how wrong I was, a seventy-foot-tall portrait of Mika is painted on the side of the parking garage at Philadelphia International Airport. She's the third dancer from the right.

we saw that an awful winter storm had blown in, I turned Mika down when she offered a bunch of us a ride home in her pickup. (Yes, the exotic journalist was also wheeling around Philly in her Hawaiian surfboard-mobile. Any argument that she was out of my league?) Two other guys happily squeezed into the truck's cab while Mika asked if I was sure.

"Yup, I'm good," I said. I trudged off into the freezing rain, hoping I didn't look as much like a dumb-ass as I felt. Sometime during that miserable hike through North Philly, it dawned on me that like all Taos, the Steve version didn't come with an endgame. Exactly how you're supposed to boomerang back from "be gone," I had no idea.

I got my answer a few days later. Mika got my number from Jen and called to thank me for the CDs. I mentioned some African food shops in West Philly she might like to check out, and soon we began spending time together. Mika was actually African American and Chinese, I learned, or maybe it was Thai? She didn't know for sure, because she'd been conceived when her mother had a brief romance in junior college with a foreign exchange student who suddenly vanished before Mika was born. Soon after, Mika's mother married her true love, an Army nurse named Dave, who took his new family with him wherever he was deployed across the country. Mika grew up in one city after another, always feeling like an outsider, never looking like anyone else, until the day they arrived in Hawaii. For the first time, people weren't constantly eyeing her long curls and cappuccino complexion and asking, "Where are you from? What *are* you?" Hawaii became her home because it treated her like family.

Mika never intended to leave Honolulu, but she decided to spend one year on the mainland while her boyfriend was learning hotel management in Hong Kong. Maybe I was still Tao-ing, but neither the boyfriend nor the ticking departure clock threw me off. Mika and I had a blast together, roaming used bookstores and attempting with tragic results to re-create stewed goat dishes that I remembered from Uganda. I told Mika about my plans to return to Africa to ride the entire continent on a motorcycle, following the fabled Capetown-to-Cairo route, and for the first time I got an inkling that we might have a future when she became genuinely intrigued about the possibility of saddling up behind me for the adventure.

Instead, we ended up in a farmhouse in West Virginia, discussing marriage. For both of us, it was a left turn we hadn't seen coming, but we could sense we were edging toward a life together and we had to figure out how it could work for real. Did it mean goodbye to Kailua beaches and *adios* to that Triumph Bonneville roaring into a Serengeti sunset? Mika had already handled the toughest jobs; she'd broken up with her boyfriend, who bolted straight for the airport and flew in from Hong Kong to try to talk her out of it, and she'd resigned herself to a longer tour of duty in Philly, where most of my freelance work was based. But before we dove in, Mika proposed a final stress test: she suggested we get out of the city and spend a week completely on our own, desert-island style, to see how happy we were without friends or a city to distract us.

She definitely picked the right spot. After driving four hours from Philly, we were rumbling down a lonely dirt road on the outskirts of Appalachia. We finally arrived at an old farmhouse hidden in the woods, miles from the nearest neighbor. We creaked open the ancient door to find a beautifully preserved gem, with cast-iron woodstoves and a greenhouse converted to a hot tub room. The first few days felt a little strange, with nowhere to go, no one to see, and not even a SEPTA bus rattling the windows to lull us to sleep at night, but by the end of the week, we felt right at home. When I found I could laze in the creek all morning and still file an article back to Philly with the cabin's creaky Internet dial-up, we were asking each other, "Why is this vacation? Why isn't this just . . . life?"

I'm a pro at tasks that require me to blow off work for no practical reason (ask me sometime how to repair vintage fountain pens), so as soon as we got home, I began hunting for a home that I knew we had no chance of buying. First I checked whether we could get

DSL in that part of West Virginia and whether the house we'd rented was even for sale. Nope, and nope. Then I spread the net wider. Mika and I began roaming the fancypants horse country between Philly and New York, pestering realtors with an insane list of requirements on the chance that maybe, somehow, there was a cheap, cozy cottage by a brook that everyone else had accidentally overlooked. We wanted something old but restored, isolated but train-accessible, rural but DSL-equipped, and, of course, compatible with a struggling writer's income, which meant the same price it sold for in 1870.

We'd gotten absolutely nowhere by the time our first daughter, Maya, was born, but instead of getting real, we kept treasure hunting. We listened to hours of Wiggles and Elmo Sing-Alongs as we hauled Maya to one death trap after another, including a burned-out shell on a soggy patch near the Delaware River with giant penises spray-painted on the few surviving walls. "For your price range," our realtor said, "this isn't bad."

For two years, we struck out. Then, late one night, something impossible popped up online. I sat there in the dark, staring at the photo and muttering, "No way." A hand-hewn log cabin on four acres, with a fieldstone chimney, a creek, its own sweet-water spring, a dirt-road walk to the Susquehanna River, and protected farmland on all sides. Only ninety minutes from downtown Philly, and it cost less per month than our apartment? Perfect!

Except . . .

"You know where this is, *riiiight*?" the realtor cautioned when I called the next morning, drawing the word out to make it clear he was in no way accountable for this folly. Only two houses were down in that hollow; everything else, for hundreds of acres around, was open farmland. Peach Bottom had no police, no local government, not even a grocery store; the only place within fifteen miles to buy food was a one-room shop in the back of an Amish farm. If we moved out there, he warned, we'd be completely on our own. That's why, tasty as it looked in photos, no one had made an offer on the place for over a year.

Whatever. It had to be better than the condemned House of Dongs, right? But during the drive out from Philly, the realtor's warning began to sink in. It felt like we were clicking off centuries rather than miles; in little more than one hour, two hundred years vanished from the landscape. McMansions and mini-malls gave way to red barns and windmills; Escalades were replaced by horse and buggies and hay wagons. We knew Lancaster County was famous for its Amish community, but we hadn't realized we were heading into an even deeper, more rustic heartland: the river hills of the "Southern End." Even by Lancaster standards, the Southern End is another world, a place where the post office has a hitching post, the kids have Drive Your Tractor to School Day and no classes on the first day of hunting season, and you're about as likely to find a gun range behind the house as a swing set.

When we located the log cabin for sale, we were stunned. Winslow Homer couldn't have done this place justice. We pulled into the dirt-and-gravel driveway to find horses grazing along the fence line, watercress blooming in the creek, and an Amish farmer rumbling past in a steel-wheeled wagon. Unbelievable. If a milkmaid had appeared with buckets of frothy cream across her shoulders, I'd have been only a little bit surprised. We threw open the car doors, raving about the gorgeous view, until I spotted something bizarre out of the corner of my eye. Or did I? When I jerked my head around to look, all I saw was the blinding afternoon sun. Weird, I thought. It looked just like—

Then it reappeared. There, stepping out of the glare, was a lone horseman watching us from the hillside across the road. He was wearing a Stetson and a serape, Josey Wales style, and had a machete and a rifle hanging from his saddle. I lifted a hand to wave, but he wheeled his horse, dug in his heels, and galloped off.

"Zach's out hunting pigs again," explained the homeowner, who'd walked down to greet us. Explanation: The thirteen-year-old who lived over the hill was riding through the cornfields, shooting groundhogs. "Don't worry, he won't shoot over here. Not unless you want him to." For the rest of the day, those brief glimpses of the Lone Pighunter and one Amish farmer were the only signs we saw of human inhabitation. If we moved here, it really would feel like volunteering for a manned mission to Mars. How would we survive when the house was snowed in and the power died? Where was the nearest hospital, and what were the schools like?

Those were all sensible questions, none of which Mika and I had ever thought to ask before pulling the realtor aside to make an offer. A few weeks later, we were leaving Philly for the Southern End.

Solitude was going to be our biggest problem—or so I thought until I heard Mika scream. I dropped the box I was unpacking in the basement and came running. I found her on the back porch, holding two-year-old Maya in her arms, edging away from a six-foot blacksnake writhing at her feet. She had been watering a row of plants, and when she stepped from one to the next, the snake dropped from the porch roof and thudded down right where Mika had been standing a second earlier.

I grabbed a shovel, thinking I could flip that monster away, but it began slithering up the porch post toward the roof. I banged the shovel blade in front of its face to drive it back; instead, the snake glided right over the shovel and disappeared into the eaves. Now what? The only thing worse than finding a giant carnivorous reptile outside your house is driving it inside, especially when inside contains a toddler you'll soon be putting to bed. I had to figure this out before dark, so I walked through the cornfield behind our house to pester, once again, our nearest neighbor, an older Amish farmer who goes by the initials "AK" (a touch of genius if he intends it as a gag, and knowing him, I can't rule it out and he'll never tell: AKA AK).

We'd been in the house only a few days, but already I'd gone to AK for advice about repairing our collapsing well, finding a desk, and buying a used chain saw, which he both sold me and taught me how to operate. When I tromped to his door with my latest problem, he didn't understand the fuss. "You're lucky," he said. "That's a rat snake. It will eat your mice and stay out of your way."

We have mice? That was a revelation. So was the fact that soon I'd be telling Mika the good news that we were now roommates with a resident serpent. "I think we'd rather have the mice," I told AK.

"Well, that's okay," he said, pushing ahead glass-half-fullishly. "It won't get *all* of them."

Before leaving, I thanked AK the way I always did, by offering my help if there was ever anything I could do for him in return. For the first time, he took me up on it. He asked me for a favor that, over the next few months, would change my life: "Can you give me a ride to the hardware store?"

That request opened my eyes to a terrific little loophole in local Amish code of conduct. Every Amish community sets its own laws: some allow foot scooters but not bicycles; others allow cars but only if they're gray or black. Down in the Southern End, the Amish are mostly Old Order, which means they can't drive cars but they can *be driven*. Way before Uber was a gleam in Garrett Camp's eye, our non-Amish neighbors had created a nice little cash economy by hiring themselves out as taxi drivers for Old Order families who needed a lift to places they couldn't reach by horse and buggy.

"Sure," I told AK, although I wasn't sure how Mika would feel about me joyriding around while she was stuck at home with a predator on the premises. AK and I hiked back across the cornfield to fetch my old Bronco. He congratulated Mika with so much charm and winking good humor on her solution to a rodent problem she didn't know she had that it convinced her—and me—that maybe the snake wasn't such a big deal after all. AK and I got into the Bronco and set off, winding along twisting farm roads until, within a few miles, we arrived at a place I never would have found on my own: behind a barn and invisible from the road was a long, white bunker. I squeezed the Bronco into a parking spot between two horse-drawn buggies, then stepped through a door into the 1800s.

Inside, the Amish hardware store was dimly lit by hissing gas lanterns. Men in black suits and straw hats searched the aisles for sheep testicle removers, hand-cranked ice-cream makers, spare wheelbarrow handles. Can't find what you need? No problem; the old gent behind the counter has an intercom consisting of a funnel attached to a long piece of plastic sink pipe stretching to another funnel at the back of the store. He'll honk a bike horn to get his storage assistant's attention, then the two of them will holler back and forth into their funnels like kids on a paper-cup phone. The whole scene was like wandering onto the set of a Henry Ford biopic, except with excellent prices on name-brand power tools.

I found a sweet deal on a splitting maul, but when I went to pay, the old gent looked at the credit card in my hand and shook his head. No one had ever bothered to post a CASH ONLY sign on the door, because anyone who could find the place already knew that (a) *hello*, no electricity means no card swipe, and (b) Old Order Amish don't do debt; they buy only what they can pay for, so they have no use for cards. I started to put the maul back, but once again, AK was there to help me out, fronting me the money without even rolling his eyes.

A few days later, I got a call from AK's son, Amos (AKA AK, JR.). When I heard what Amos wanted, I was psyched. I'd been stuck at my desk in the basement all morning, working on an overdue magazine assignment and dreading a call from my editor, so I was dying for any excuse to escape. Soon Amos and I were in the Bronco, heading for another behind-the-barn hot spot. This time he directed me to a farm less than two miles from the house. No sign was posted out front, but those in the know were aware that the back shed was actually a semipro butcher shop. Legally, they couldn't sell meat, but you could rent their skill, so if you brought in your own animal, they'd kill it and send you home with bins full of steaks, chops, sausage, jerky, and bologna to fill your basement freezer chest.

Amos lucked out; he'd wanted some free bones for his dogs, and the butcher boys hooked him up with two trash cans full of bloody carcasses. During the drive home, the weirdness of the moment sank in: how, out of all life's paths, did I end up cruising around with a trunk full of discarded cow and a guy who speaks ancient German and doesn't believe in zippers? Amos and I began trading stories, and it turned out we had a lot more in common than I'd realized. He had just turned thirty, and like me, he was still adjusting to life as a first-time dad and homeowner. He told great stories, especially about his younger brother, who canoed all the way across the Susquehanna River in the middle of a winter squall to visit his girlfriend. Amos and I had such a good time that we made plans to get together again the next morning to cut firewood.

From then on, Amos was my jungle guide to the Southern End. Every few days, he'd call with another adventure in mind, and I'd immediately slap the laptop shut and head out the door. Amos took me to my first "mud sale," a local firehouse fund-raiser held each spring thaw, and introduced me to the Tuesday night poultry auction, where I accidentally bought seventeen massive roosters instead of hens. Amos was a wizard at sensing when the power company was taking down trees on remote back roads, and since we both depended on wood to heat our homes, we'd speed off together to chain-saw the logs and heave them into my truck before anyone else got to them.

One freezing February afternoon, Amos called with an urgent tip: his older brother had three big boars he wanted to get off the property, so if we butchered them that night, we'd get a ton of cheap, fresh pork. Mika's family was visiting, and to this day, I can't figure out why I thought her dad would enjoy this. Long past midnight, the poor guy was shivering and up to his elbows in gore, trying to avoid hacking off his own frozen fingers as we helped Amos dismember the massive animals with handsaws in an icy barn lit only by headlamps. It was nearly dawn when we got home, bloodstained and filthy, looking like *Lord of the Flies* kids who never made it off the island.

Mika became friends with Amos's wife, Katie, who taught her that sheep's milk was a great alternative for the dairy-intolerant.

Mika with her ewe

Mika began making cheese with Katie, and would drive her to her midwife appointments as Amos and Katie's family grew to five children. When we had our own second daughter, our kids would color and play board games when we got together for family meals. One evening, I was out for a run past AK's house when he waved me over. Did I want some raw cream left over from the milk pickup? I could carry the cream in two half-empty jars, he suggested, so while I was running, it would slosh into butter. While we were talking, a jubilant Amos showed up. He'd shot three deer while bowhunting and had plenty to share. My hands were full, so the only option was to shove the haunch he offered me into the back of my running shorts. When I got home, the girls were sitting down to dinner as I came through the door with blood oozing down my legs and two jars of clumpy gook, but by then, they were too Southern End to be surprised anymore.

Personally, I felt my own identity clicking over from city to country the day three strangers passed our house and I thought, *Who the hell are those weirdos?* They didn't greet me with a quiet wave, like the Amish do, and they weren't driving a 4x4 pickup, like everyone else. They were clopping along in some sort of big-wheeled cart pulled by two well-groomed horses, the kind of thing Queen Elizabeth might use to tour Buckingham Palace's back forty.

"*Hallooooo,*" hollered the woman at the reins, waving a long whip. One guy sat beside her on the bench, and the other balanced on the rear fender like a footman. They yelled and waved too. If they hadn't spotted me already, I'd have ducked out of sight. Instead, I mouthed "Hey" and gave the least perceptible nod I could get away with, then faded back into the house. For months afterward, I kept crossing paths with them, usually when I was out for a run on the dirt road to the river and they were tooling around in their carriage. It was always all three of them, Whip Woman and her two buds, and there was something about their jolliness and sophisticated horsemanship that made me wary. I didn't know how they fit into Southern End life. They seemed like such . . . outsiders.

I mostly managed to keep my distance, until the day we ran smack into them in the woods. Mika and the girls and I were hiking the nearby Conestoga Trail with some friends, fighting our way up stony hills known as the "toughest ten miles east of the Rockies." When we made it back to the trailhead, we heard a clatter of hooves behind us. We got out of the way just as the two carriage guys burst from the trees on horseback. Behind them was Whip Woman riding a donkey. I'd never seen anyone on a donkey before—no one, at least, who wasn't playing the Madonna or a Mexican bandito—so despite my doubts about this crew, we couldn't resist crowding around for a closer look.

"My ass is breaking my *ass*," Whip Woman hooted. She slid down from the saddle and swatted her backside. "The Southern End is taking a *beating* today." The kids were enchanted by the donkey, and I had to admit I was charmed at the way Whip Woman not only cut straight to the butt jokes, but also introduced Muf-

Tanya McKean, best donkey trainer in the Southern End

fin before even mentioning her own name and then immediately bustled over to show the shyest girl how to shove a fistful of horse treats into Muffin's mouth. "Good job, girlfriend!" she cheered.

"I'm Tanya," she finally got around to telling us. The two guys were her husband, Scott, and their horse-show partner, Paul. Right around the time I first saw them, Tanya and Scott had relocated here from outside the county and bought a small farm that was even harder to find than ours. Paul was an engineer who lived three hours away, but nearly every weekend he drove down to work out his horse and hang with Scott and Tanya. None of them had kids; instead, they were devoted to the fine art of carriage competition and, in Tanya's case, to nearly every critter she encountered. Tanya was the adopted child of sweet but stiff British parents, and she learned early on that with animals, she didn't have to throttle back her bubbling natural warmth. Dogs don't freeze up when you hug them; horses give you the same oxytocin rush of love and appreciation no matter how much you brush them. Luckily, her parents indulged her; from the time she was eleven, Tanya had a horse of her own and went on to study equine science in college and veterinary tech school.

By now, I was feeling like an idiot for having avoided Tanya and her crew for so long. Mostly because they were so fun, but partly (okay, equally) because Muffin was such a star. Up close, donkeys just ooze charisma. Yes, they look ridiculous with those stubby bodies and Bugs Bunny ears, but that just makes their Latin lover eyes all the more heartwarming. While we were talking, the horses were pacing and pawing the ground, but Muffin stood quietly, holding our gaze as if she genuinely, *soulfully*, wanted to connect. Or con us out of our granola bars. Either way, Muffin was definitely next level over those horses. We were all sad to see her go when it was time for Tanya and the fellas to load their rides back into the trailer and leave.

On the drive home, Mika and the girls and I cut loose with one of those conversations that are such a blast because you can pretend it's real even though you know, deep down, none of this stuff is ever going to happen. My youngest daughter, Sophie, was the one who really went wild. All she wanted for her tenth birthday, she said, was a donkey like Muffin. We could keep it behind the house, so whenever she wanted, she could just saddle up and ride through the hayfields. *Wait a sec* . . . she could even ride to *school*, and I could run up later and bring the donkey home!

Sure, why not? Someday, we really ought to get a donkey of our own.

Someday.

4

A You Operation

"I can't believe he isn't dead," an appalled friend who raises sheep in upstate New York replied when I sent her a photo of Sherman. "I have seen farmers here put down animals in far better shape."

After Scott's hacksaw surgery and Tanya's tough-love shearing, we left Sherman in peace for the rest of the day. We held our breath, waiting to see if he could walk, but he spent all day pressed against the side of our little brown barn, head hanging, looking like he'd been led there for execution. Whether he was frightened, or in pain, or just confused, we still couldn't tell.

But now it was getting dark, which meant things were about to get ugly.

I'd moved our little herd of goats and sheep to another meadow for the day, figuring Sherman would settle in more easily if he could explore his new home on his own. Once the sun began to set, the sheep began *baa*ing by the gate, eager to get back into the shed for the night. I didn't know how they'd react if they rushed in to find this scruffy-looking stranger in their yard, and the last thing Sherman needed was any kind of physical challenge. Our animals are all pretty gentle, but we have a ram and two billy goats who are, by exact scientific standards, a bunch of boneheads. Their normal greeting for newcomers is to rear up on their hind legs and race straight at them until they knock noggins. It's not malicious; I've seen enough of their headbanging to realize it's just their way of high-fiving. Still, playful or not, no way was I letting them anywhere near Sherman tonight.

Sliding them past Sherman without one of them breaking loose was going to be tricky. My hope was that if I waited till dark, they'd all be in such a hurry to get inside that they'd rush right past Sherman without stopping. Overnight, they could get used to his smell and (fingers crossed) be adjusted to the presence of this weird new creature by sunrise. It wasn't much of a plan, but it turned out the sheep and goats were down with it. When I opened the gate, they all breezed past Sherman and were heading straight for the stalls.

Until suddenly, one of them froze in his tracks. I peered through the dark, hoping it wasn't. . . .

Yup. Here comes Lawrence.

Every problem I'd solved by getting rid of Bamboozle and Skeedaddle, I un-solved by getting Lawrence. Yes, I realize now it was a mistake, but tell me you wouldn't have done the same thing: One Saturday morning in March, I drove to the farm of our Amish friend Elam to drop off the ram he'd lent us to—as the Amish say—"freshen" our two ewes. Over the years, we'd developed a little swapping circle of local sheep raisers who trade studs every year to prevent in-breeding in our herds. Elam's ram was known for fathering twins, so we could expect to have four new critters bounding around the meadow within a few months. That meant that no matter what kind of adorable little creature Elam had for sale, we didn't have the room for any more mouths. Returning the ram was strictly a drop-off operation, Mika reminded me.

"Absolutely right," I agreed.

When I got to Elam's, it took a while for me finally to track him down in the back of the barn. He was gazing over the half-door of a stall.

"Aren't they beautiful?" he said. "When I heard what they were, I couldn't resist."

Elam was a regular at the livestock auction held every Friday night at the Green Dragon Farmers Market on the outskirts of Lancaster. He was always coming back with great finds, like Katahdin sheep, which shed hair like dogs instead of being sheared, and Tennessee Fainting Goats, which freeze into living statues whenever they're scared. Once, he even got a llama that acts like a watchdog and protects his herds from predators. But this time, had he actually found a pair of—

"Gazelles?" I blurted. "You got *gazelles*?"

"They look like that, right?" Elam said. "They're actually Oberhasli. They're a kind of Alpine goat, really hard to find around here. They're famous for being amazing milkers and easy to work with."

Inside the stall were a mother goat and her young kid. Both were a beautiful chocolate brown, with oddly wide-veering horns and a black streak down their spines, which made them look noble and wild, like they'd just wandered in from the African savannah.

"What's wrong with the baby?" I asked. The top half of its ears were missing, like they'd been snipped off with scissors.

"Remember that cold snap a few weeks back? The owner said he was born late at night and they didn't find him till morning. The mother kept all of him warm except the tips of his ears, and they just dropped off from frostbite."

I spent the drive back home trying out my excuses: "Look what I found wandering in the street! . . . I think Elam was going to eat them . . . No, they're going to *make* us money." Just before I pulled into the driveway, I realized I was holding a winning card that didn't require lying; the two brown goats were so cute, my best move might be to keep my mouth shut until Mika got a look at them. Luckily, it worked; the lop-eared baby was a heart melter who won over my wife and daughters the second they laid eyes on him, and Mama Oberhasli turned out to be sweet-natured and an amazingly productive milker. But we weren't going to take any risks; the baby was a boy, meaning it was prone to mischief, and we didn't want to end up with another Bamboozle on our hands by tagging it with a name that could release its inner demons, something like Beetlejuice or McConaughey. The safest bet, we decided, was to call him Lawrence, after the gentle, nerdy fourth-grader who played keyboards in *School of Rock*.

Perfect. Baggage-free.

Soon we were being awakened at five in the morning by screeching brakes followed by irate neighbors hammering on the back door to say they'd nearly hit a furry doofus in the middle of the road. It had taken only a few weeks for Lawrence to grow into a lean, lanky-legged youngster with a body aerodynamically designed for maximum lift and glide. I would scramble out the door, blurry-eyed and apologetic, and hustle off in bare feet to chase Lawrence back behind the fence. At first, I blamed the name—after all, the original Lawrence *did* join Jack Black's rebellion and got all chesty with drummer Spazzy McGee—but then I did some due diligence in *Modern Farmer* and got the good news about Oberhaslis:

> *The dogs of the goat world, they are quiet and companionable.*

And the bad:

> *But that doesn't mean they won't go on the lam. With powerful hind legs, an Oberhasli can easily bound over a fence or to the top of a car.*

Lawrence, in other words, was genetically wired to crave cuddles and elude anything that tried to prevent them. During thunderstorms, we'd be startled by a sudden pounding on the door before remembering that it was just Lawrence, feeling lonely and wanting to come inside. We got a lucky break when our best Amish friends, Katie and Amos, came up with the idea of breeding Lawrence with their Boer goats to create a super-herd of friendly, meaty milkers.

Traveling with Lawrence was easy; I just clicked a leash on his collar and strolled the half mile like I was walking a dog. I dropped him off with a mob of Amos's nanny goats and said good-bye as he ran off happily to meet the ladies. Two weeks later, a tractor trundled into the driveway with a dog crate on the back. Inside was Lawrence, returned in disgrace. Even though Amos's fence was higher than ours and was reinforced with an electric shock wire across the top, it was no match for those Oberhasli springs. It was bad enough that Katie had to constantly race out the door to shoo Lawrence out of her garden, but the final straw was the phone book. The Amish keep their telephones out by the barn, usually in a little shed that looks like an outhouse. Early one morning, Katie went out to make a call and found Lawrence inside, munching the thick notebook containing the family's lifetime collection of phone numbers.

No matter how aggravating Lawrence was, though, I couldn't stay mad for long. I would come boiling out the door at dawn to retrieve him from the road, only to find him galumphing toward me as if he'd been looking forward to seeing me all morning. To him, breakouts were less about freedom and more about friendship. Big and strong as he'd become, in his heart Lawrence was still the frozen-eared little furball shivering next to his mom on his first night in the world.

Lawrence, as usual, was bringing up the rear as the herd came in for the night, always the last kid to leave the playground when it got dark. Suddenly he pivoted, head high and on alert. He sniffed the air, zeroing in on the weird, shaggy shape standing motionless against the wall of the barn. Sherman was cornered. Even if he suddenly found the power to move, he had no room to escape whatever was going to happen next.

Chummy as he was, Lawrence was the last one I wanted anywhere near a dazed, traumatized donkey whose badly damaged feet and months of captivity in a tiny shed left him practically

Lawrence

immobile, a sick and sitting target. Lawrence's bounding welcome tended to freak out children and other animals meeting him for the first time, mostly because they had no idea why this galloping madman with two giant bone spears on his head was charging at them, full speed. There was no way I could get to Lawrence in time without the risk of spooking him and making things worse, so I braced myself and hoped Lawrence would change his mind and hurry off once he noticed the herd had left him behind. Instead, he did something I'd rarely seen him do before: he took his time.

Lawrence approached Sherman slowly, cautiously, as if he were tiptoeing. It's so weird to see him so nervous, I thought, until I realized something else was going on: When Lawrence got close to Sherman, he shoved his nose right up against the donkey's flank and began sniffing him, curious but careful, working from head to toe. Sherman didn't move a muscle, not even flinching when the points of Lawrence's horns were in his face.

I watched, edgy and ready to spring into action the second that goofy goat gave a hint of feistiness. Lawrence kept snuffling along, working his way all the way down Sherman's flank and into the high-impact zone behind his tail, circling Sherman until he disappeared behind his far side. He finally reemerged by Sherman's drooping head, his inspection tour complete. Whatever story Lawrence's nose picked up must have troubled him, because he then did something that, in an instant, made up for every flowerbed he had ever ruined, every driver he had ever terrified, every torn fence I've had to replace.

Lawrence lay down beside the sick donkey, curled his legs beneath him, and settled in for the night.

When I came out the next morning, Lawrence was still there. It was baffling. Sunrise *never* found Lawrence anywhere near where he'd been the night before; ordinarily he'd be messing around with the other animals by now, then hurrying down to the front gate to see if the kids who caught the school bus at the end of our driveway were in the mood to share a little treat from their lunch boxes or at least scratch his ears.

But not today. Nothing could make him abandon this sick stranger. Nothing—

Except breakfast.

I pulled open the door of the hay stall and suddenly Lawrence, that chowhound, scrambled to his feet. He sprinted toward me, shoving his way into the middle of the scrum of sheep and goats that were already crowding around. I shoved a half bale of hay into the hayrack and got out of the way, leaving the critters to tear into their meal. Behind me, I sensed something approaching. I turned and found Sherman taking a few slow, tentative steps. He stopped a few paces away from the hay feeder, keeping himself clear of the mayhem, but as Lawrence shifted to the left, so did

Sherman. When Lawrence spotted a better place to attack the food and moved to the right, Sherman quietly followed. For the rest of the morning, Sherman kept his distance from the other animals. But anytime Lawrence made a move, a long-eared shadow was right behind.

I sat on the grass, watching Lawrence fill his belly and then get started on his full docket of daily shenanigans. He got a running start and began butting heads with big old Buddy, the lovable gelding ram we'd raised from the time he was a bottle-fed baby. Buddy tolerated Lawrence for a while and then put an end to his nonsense, bulling Lawrence back to the fence and sending him scampering off for someone smaller to mess with. About forty yards away, he spotted the perfect target: Chili Dog, our other billy goat, was grazing quietly, clueless that behind him, a trouble-seeking missile was locking in. Lawrence began ambling off in his direction—and marching along behind him, slow but determined, came Sherman.

That made up my mind. Okay, I thought. Time to give this a try.

I hustled to the truck and set off for the feed store. Tanya was coming by later to check up on her patient. If I really hurried, I might have just enough time to put this scheme to a test before she saw what I was up to.

Tanya rolled up early in the afternoon. "Hiya, Sherman," she hollered, her voice ringing with affection as she climbed out of the truck. "How are you, little man?" To me, her tone was a little more grave. "He's going to need a sheath cleaning right away."

"No, he didn't come with one," I said, figuring that must be some kind of horse blanket. "All he had was that crappy halter."

"I guarantee you, he's got a sheath," Tanya said. "I'm looking at it right now. It's his penis."

"We've got to clean his penis?"

"Not we. *You.* This is a 'you' operation. You better learn now because if he lives, you'll be doing it every three or four months."

"Seriously?" I'd devoured horse books as a kid, and nowhere in the *Misty of Chincoteague* saga was there any mention of Grandpa Clarence fiddling around with any of the colts' junk. Tanya was dead serious, though: because Sherman was neutered, his dormant penis was susceptible to waxy buildup that could cause serious issues. The only way to get the job done is to ease your fingers into the donkey's abdomen, pull down the retracted penis, and swab it carefully with warm, soapy water while trying not to get kicked into outer space.

"Last thing, you stick a finger in the hole—" Tanya continued.

"Of what? The penis?"

"Right up in there. You've got to pull out the wax ball. That's the 'bean.' The bean is a killer. Totally wrecks his bladder." Apparently if Sherman had a new life ahead, so would I: every three months, I'd be reminding myself to file quarterly taxes and degrease my donkey's downspout.

Now that she was finished terrorizing me, Tanya was ready for her patient. "Let's get a look at you, Shermie," Tanya called as we came through the gate.

Sherman's ears flicked up. I don't know if he remembered his name, but oh, boy, did he remember Tanya. He didn't lift his head, but his ears stayed on high alert as he pivoted and suddenly, for the first time, seemed very interested in the sheep. He trudged toward them and kept going, plowing into the middle of the herd. Lawrence spotted his new buddy on the move and ran over, shoving in until they were both surrounded by sheep. It wasn't much of a hiding spot, but I was impressed with Sherman for trying.

"Look at him walking!" Tanya said, not a bit offended. "And he's got a friend! Dude, this is amazing."

She followed Sherman into the scrum. She reached out to pet Sherman's mane, but his head bobbed and she got only air. She tried and missed again, as Sherman backed deeper into the herd. As an exercise in sheer irritation, it was a joy to behold; Sherman taking evasive maneuvers was like watching Jackie Chan slip punches. Whatever that Crazy Clipper Woman had in mind for today, Sherman was having none of it.

"Someone remembered he's a donkey," Tanya said. "I love it."

We finally got hold of Sherman's halter, then pushed pesky Lawrence out of the way. Tanya took a syringe of antibiotics from her jacket pocket, bit the cap off with her teeth, and pinched a fold of Sherman's hide back near his haunch. With one swift move, she injected him so deftly that Sherman didn't seem to even feel the needle. Tanya dug into her other pocket and pulled out a tube of deworming paste to treat Sherman's intestinal parasites. "He's going to like this. It tastes like fresh apples," Tanya said. "So let's use it as a distraction." Tanya instructed me to open the tube and let Sherman take a big whiff of Golden Delicious goodness before squirting a dab between his teeth. Tanya kept her eye on him, and the second Sherman began stretching his mouth toward the paste, she pounced: she eased in a rectal thermometer and, in a masterpiece of pain-free nursing, managed to take Sherman's temperature before he noticed what was going on.

"Little hot," she decided, and reached into yet another pocket of her mobile-pharmacy outerwear for a tube of Banamine, an analgesic paste that should bring down Sherman's fever and help ease any muscular pain that could contract his intestines. "That should make him feel a lot better," she said. "It's like donkey Tylenol." Overall, Tanya was cautiously optimistic. It could be weeks before we'd know if Sherman's gut was in order and his bloodstream was free of infection, Tanya guessed. But the big thing was his feet, and they were now looking hopeful.

"He's moving well," Tanya said. "A lot better than I expected."

"Yeah, but he won't set foot on anything hard," I said. "He won't even step on gravel in the driveway."

"Now why would you want him to?" she asked, her tone expressing everything she wasn't saying out loud: *I leave my sick patient alone with you for one night, and already you're screwing things up?*

"I had this idea," I began, not enjoying the way the word "idea" made Tanya's skeptical eyebrow shoot up even higher. "Everything you said yesterday about despair, I took seriously. And here's what I came up with . . ."

5

Miners, Muckers, and Mean Mother—

Ten years earlier, I'd gone to Leadville, Colorado, an old mining town high in the Rockies, in search of ghosts.

I was there to find out what really happened back in the 1990s when a strange band of men in skirts and sandals suddenly appeared to compete in Leadville's famous 100-mile footrace—and just as abruptly disappeared. They were Tarahumara Indians, members of a reclusive tribe who live in the depths of Mexico's remote Copper Canyon. For centuries, all kinds of tall tales had drifted out of the canyons about the extraordinary running ability of the Tarahumara. Few outsiders had ever seen them in action, but those who did claimed that the Tarahumara were near-superhuman athletes who could scamper from peak to peak faster than a horse and knock off ten marathons in a row (yes, that's more than 260 miles) without stopping.

The Tarahumara almost never emerge from the canyons to compete, so when about a dozen of them showed up in Leadville in 1993, they created a sensation. Despite their reputation, it was hard not to feel sorry for these shy, uncertain guys as they approached the starting line, wearing huaraches they'd made the day before from junkyard tires—until the gun blasted and they proceeded to kick the living daylights out of hundreds of younger, better-equipped runners. A fifty-five-year-old Tarahumara farmer finished first, followed close behind by seven other Tarahumara villagers in the top ten. The next year, the villagers returned and crushed Leadville for the second time. And then they were gone, never to return.

So what happened? The mystery of the Tarahumara that brought me to Leadville soon launched me on the crazy adventure that I describe in my book *Born to Run*. I'd gone to Leadville braced to butt heads with tough old miners who didn't appreciate reporters sniffing around with a lot of nosy questions, and it took about four minutes for those miners to show me that what I didn't know about Leadville was a lot. My first morning in town, I sat face-to-face for two hours with Ken Chlouber, the pit chief who created the Leadville 100 as a way to save his town after the mine closed. Ken answered every question I threw at him until he got tired of sitting around. "Hey, let's go snowshoeing," he suggested.

Soon after, we were tromping through knee-high drifts, climbing into the Rockies so I could see for myself where the Tarahumara had triumphed. Ken was nearly seventy years old but charged up the mountain with such savage power that I was almost too winded to ask him to slow down. Spots were spinning in front of my eyes, and I was sucking air like a drowning swimmer. Leadville is a mile closer to the sun than Mile High Denver, making it the highest city in the continental United States. The air is so thin that first-time visitors often get headaches as soon as they step out of their car.

"Worth it, though, right?" Ken said. I lifted my drooping head and took in the herd of elk grazing down below, the glitter from the snaking Arkansas River, the breathtaking beauty of the powdery spruce forest soaring toward the peak above.

"Yeah, it's really—" I began, but Ken was suddenly struck by inspiration.

"You've got to come back this summer!" he blurted. "For Boom Day. First time I saw Boom Day, I knew I was here to stay."

Ken had grown up on a farm in Shawnee, Oklahoma, and wandered into Leadville as he drifted from job to job. The mines paid pretty well, but what hooked him for good was the morning in August when he woke up to find a stampede of animals thundering down Leadville's main street. A herd of donkeys was galloping wildly while a gang of Ken's fellow miners followed close behind, running their hearts out and holding on as best they could to the donkeys' halter ropes. The mob rounded the corner and was gone, charging up the dirt road for the thirteen-mile climb to Mosquito Pass.

Ken had just gotten his first glimpse of burro racing, and he wouldn't need a second. (In case you were wondering, "burro" is just the Spanish word for "donkey." Much like "pop" versus "soda," whichever term you use depends on whether you're east, west, or on top of the Rockies.)

Pack burro racing was a throwback to the Gold Rush days, Ken was told, back when prospectors would hit pay dirt, heave their gear onto their burros, and hightail it to town to file their claims. By 1915, the old prospectors had disappeared from the mountains, but the burros remained. Mining had gone deep underground, shifting from precious metals to industrial minerals, and those compact, steady-tempered burros, which wouldn't go berserk at the blast of a dynamite cap, were still needed to haul the heavy ore wagons to the surface. The miners who worked beside them in the darkness bonded with the burros and treated them like personal pets. On weekends, they'd bring their kids to the big corrals outside of town to feed apples to their buddies over the fence. Eventually, they said to hell with the fence and began busting the burros out, taking them on day hikes into the mountains.

You can see where this is going, right? There is no way you can combine a bunch of suddenly freed animals with a gang of "miners, muckers, and mean motherjumpers," as Ken Chlouber affectionately calls them, without it turning into some kind of redneck rodeo. By the 1940s, miners were racing one another over the trails and right into downtown Leadville. It got to the point where it was weird to walk into the Silver Dollar Saloon on a Saturday and *not* find a donkey standing at the bar alongside a thirsty two-legged teammate.

Over time, the distances got longer, the miners got faster, and the bets got bigger until, in 1949, an epic challenge was thrown down: Anyone foolish enough to try was invited to a twenty-three-mile, all-comers burro race stretching from the Silver Dollar, up and over a 13,500-foot mountain, and back down the far side to the Prunes memorial in Fairplay, erected in honor of a donkey who wandered around Fairplay for years as the town's shared pet. Get to Prunes first, and you walk away with $500. Assuming you still can.

Edna Miller was awed by the race, a little less by the racers. She watched the miners straggle into Fairplay, some so exhausted they couldn't speak, and figured those boys weren't doing anything a woman couldn't handle. Especially if it meant a $500 payday.

Hold on now, Edna demanded of the miners and muckers. Why can't women run?

The miners and muckers looked at one another and shrugged. Who said they can't?

Well, how about the International Olympic Committee, the Amateur Athletic Union, and the American Medical Association? For decades *after* 1949, physicians were still spouting the theory that too much exertion would make a woman's uterus and ovaries shake loose. Or explode: As recently as 2010, the president of the International Ski Federation was explaining that ski jumping was "not appropriate for ladies from a medical point of view" because, of course, it was well known that the uterus could "burst" upon landing. Naturally, the ladies couldn't be trusted to take care of their own organs, so men stepped in. Until 1980, women were banned from any Olympic track event longer than 800 meters: not a 10K, not a 5K, but a 0.8K—less than half a mile. Meanwhile, in

Boston, Running While Female was literally a crime: any woman who dared attempt the Boston Marathon in the 1960s was subject to arrest by the cops or, if your dad was in charge, a beating. "If that girl were my daughter, I would spank her," race director Will Cloney famously snarled after Kathrine Switzer finagled her way onto the course in 1967.

But in Leadville, the hardrock miners saw things a little differently.

"Out West, we've always known that women were cut from the same leather as men," said Curtis Imrie, the legendary burro whisperer and three-time world champion who was happy to talk about the many times he'd been smoked by women like Barb Dolan and Karen Thorpe. "Burro racing has none of that nonsense you have back East about 'protecting' women."

In 1951, the miners and muckers not only welcomed Edna Miller to the starting line but encouraged her to bring some pals. Within four years, women were all over the mountain, making up a quarter of the entire burro-racing field. Looking back, it's kind of insane that in Boston, grouchy old men with cigars and overcoats would keep declaring until 1972 that women were too dainty to run their marathon, while in Colorado, the "ladies" had been tearing up a far more grueling challenge for twenty years. Boston likes to boast that it's America's oldest marathon, but that's true for only *some* Americans. For the other half of the population, the ones who were outlawed for decades from even entering, it's as if the race didn't exist.

So for all Americans, men and women alike, our oldest marathon is the one that's always been open to everyone. It's not going to cost you a fortune, and you don't have to qualify. All you have to do is show up, borrow a donkey, and get ready for battle.

"First-timers either love it and never stop, or disappear and never return," Ken Chlouber told me. "You'll either be cured or addicted."

Within a month of seeing his first race, Ken had rented a bit of pasture and bought Mork, a burro so big it could look Ken dead in the eye and, as Ken soon discovered, kick him right in the chest. Ken would leave the mines at daybreak after working all night and go straight off to train with his new partner, only to limp home an hour later, bruised and confused. Nothing about these creatures made any sense. Ken had grown up in the saddle as a competitive bull rider, so getting the burro under control was supposed to be the easy part. *Running* was supposed to be the pain; Ken hated moving under his own power at any speed faster than a saunter. But with Mork, all that was flipped around. On good days, Ken found he loved trotting through the pine forest with another creature, both of them working hard and panting together in perfect sync.

But on the bad days . . .

"If you and that burro aren't of the same opinion about where you're going and how fast, it can drag you up the side of a cliff or through a boulder field," Ken warned me. "And there ain't nothing you can do but hold on and holler."

Ken was committed, though. Once he started racing, he couldn't stop. Over the next forty years, Ken and Leadville would both endure a series of tragedies, triumphs, and transformations. Ken first arrived as a stranger, a struggling furniture salesman trying to support a wife and a new baby, and went on to become a rock-blasting crew boss, then an unemployed miner, and, finally, a local legend who saved his adopted home. After the Climax Mine shut down, taking nearly every job in town with it, Leadville was close to death. That's when Ken came up with the genius idea of staging a 100-mile footrace. Leadville was reborn as an off-brand Aspen with a vibrant adventure-sport economy, and Ken was on his way to the capital as a congressman and state senator. But no matter how much things changed, one thing never varied: Every year, the burros of Leadville took their place at the starting line, and every single year, Ken Chlouber was right there with them. Even when he had to saw the cast off the leg he'd broken in one race so he could run in the next one.

When I came back to Leadville that August (yeah, he talked me into it), I shook Ken's hand and found myself gripping bandages and a splint. "Left half of me on the rocks last week," Ken said with a shrug. He'd been running a twenty-nine-mile race in nearby Fairplay when his burro got overexcited and barreled over him, tumbling the two of them down the trail. Ken broke three fingers and hamburgered his legs, but he still got to his feet, chased down his burro, and continued on to the finish line. He was sixty-eight years old.

Despite his crushed hand, Ken couldn't wait to get me out with him on the Leadville course. We headed down Harrison Avenue, Leadville's main thoroughfare, to a parking lot behind the bank where Ken had left his stock trailer. We still had a good hour till the starting gun, but already tension was in the air. Burros were pacing and twisting nervously in place, held tight by racers trying to keep them clear of the spectators who were already crowding the curb.

The basics, Ken explained, couldn't be more basic: You can use any size burro, from a mini to a Mammoth, but no mules! (Mules are half horse.) You and your burro run up the side of a mountain and back down again, in this case for a total of twenty-six miles. You can't ride, and your burro has to carry a thirty-three-pound packsaddle with a prospector's traditional tools: a pan, shovel, and pick. If your burro busts loose, you have to catch it and bring it back to where you lost it before you can continue racing. That can add miles to your day if you're running with an "uncut Jack"—a sexually active male, which will be fast and strong but prone to nuttiness. Jacks are known for fighting, bolting after female burros, or just smelling something weird and tearing off into the woods.

"Hey ya, Hal!" Ken shouted to a lean, sun-browned guy wearing battered old Carhartts and stepping out of a pickup. To me, he said, "That's Hal Walter. Very tough. Very smart. Always has a really good animal." Hal was so good, he turned pro. The sport boomed for a while in the 1980s, with enough small towns in Colorado and Arizona hosting races that Hal could barnstorm the back roads with his burro in a trailer and make a living off his prize money. We walked over to say hi, but Hal seemed a million miles away. He barely made eye contact, looking past us and mumbling something about snowballs before wandering off.

"Is there still snow on the trail?" I asked.

"He means 'Sobal,'" Ken explained as we walked on. "He's wondering if Tom Sobal is here. Tom's one of the few guys who can put Hal away." Hal is ordinarily a friendly guy who loves to talk burros, Ken said, but even though he's one of the greatest champions in the sport, he still gets jittery before a race. "Don't let it throw you," Ken advised.

We reached Ken's trailer, and he threw open the back doors. He whistled, and two eager burros trotted out. Ken tied them to a Stop sign, then went back to the trailer. Deep inside, something was still thumping around. Ken whistled again, then cursed. "Time for the Persuader," he said. He pulled a two-by-four from his pickup truck and slid it through the trailer's side slats, leaning back hard as he tried sproinging the last animal out using a fulcrum and lever. Nope.

Ken recruited two spectators, then another, until five of us with a combined body weight of half a ton were hauling on the burro's rope in a tug-of-war. No dice. Pissed off now, Ken hitched the rope to his pickup and threw it into four-wheel drive. One balking step at a time, a nearly six-foot donkey slowly emerged.

Ken handed me the rope. "This one's yours."

Race time was coming up fast. "Get your children back!" the announcer shouted. "These animals have been known to tear into the crowd and wreak bloody havoc." Parents pushed their kids behind them, then took a closer look at the size of the burros compared to the size of the racers trying to hold them, and backed away themselves.

"He's called Blue Note," Ken shouted. *Bleh bleh bleh*, he continued, but I lost it in the crowd noise and the fact that he'd just said

He. Great—a Jack. I gripped Blue Note even tighter by the halter, trying to stop him from turning around and around in anxious circles. I was getting dizzy from all the spinning when I heard the announcer shout, "TEN!"

The crowd picked up the count—"NINE! EIGHT!"—while Ken tried to give me one last bit of advice. "There are two ways to start a burro race," he said. "You can toss a cap in the air, nice and calm."

"FOUR . . . THREE . . ."

"Then there's our way," Ken concluded.

Leadville's mayor let 'er rip with both shotgun barrels, and Harrison Avenue turned into Pamplona. Burros erupted, with Blue Note and me somewhere in the middle. I tried to yank Blue Note out of sprint pace, but he was too wired by the shrieking crowd and clattering hooves. It was all I could do to hang on to the rope as we made the turn off Harrison Avenue and began the steep climb up Seventh Street.

Suddenly, I heard a scream—not cheering, but a full-on horror-movie shriek of terror. I jerked my head around to see Curtis Imrie, the master burro trainer who'd raced every year since 1974, being

Curtis Imrie fights his mammoth racing burro in the Leadville Boom Day race.

dragged on his back by the biggest donkey I'd ever seen. Curtis's leg was tangled in the lead rope and he was trying to kick free, but the more he struggled, the more the frantic animal tried to get away. I started to let go of my rope, then clutched it again. What would Blue Note do if I ran to Curtis's rescue? Wouldn't he plunge into the kids and grannies on the sidewalk?

Ken's son, Cole, leaped onto his mountain bike and charged after Curtis. I lost sight of them as Blue Note hammered on. I was dying for air, too dizzy to think anything besides *You gotta hang on. Gotta. Gotta* . . . Suddenly, I ran smack into Blue Note's butt. He'd stopped in the middle of the road. Another runner had halted his burro by wrapping the rope around a Stop sign, and Blue Note joined them. The other runner and I dropped our hands to our knees, sucking air, while the two burros watched us. I'd survived three minutes of Leadville's burro race. Four hours, fifty-seven minutes to go.

I was still recovering when I heard hoofbeats behind us and there, unbelievably, was old Curtis, not only back on his feet but back in the race. "I hate losing blood this early," he grumbled as he stopped to see if we were okay. The three of us headed off together, climbing the dirt logging road through a tunnel of whispering junipers. At Mile 11 or so, I began to think I just might pull this off. That's when Blue Note hit the brakes again.

"Hey ya!" I shouted, as Curtis and the other runner trotted on. "Hey ya!" I tugged on Blue Note's rope, then grabbed his halter with both hands and leaned back on my heels. Nothing. A spectator hiked up to lend a hand. "The only time I tried this, I gave up," he said. "I tied the burro to a tree and ran down on my own."

He pulled Blue Note's head while I pushed, then we switched. Hal Walter and Tom Sobal flew past us, heading downhill on the home leg. Runner after runner soon followed, all of them shouting tips and encouragement. The Mosquito Pass turnaround was so close I could almost see it, but Blue Note wouldn't budge. Half an hour later, I was in the same place when Ken and Curtis came jogging down in last place.

"Bring the sumbitch home," Ken panted. "He's quitting, not you."

Fair enough. I spun Blue Note around, and this time he moved—sort of. He ran a few yards, stopped to graze, started again. Somehow we'd swapped roles: I'd become the pack animal, towing deadweight. Five hours after the starting gun, I finally made it back to Harrison Avenue.

"Not for lack of strength or determination," shouted Ken, who'd waited around to cheer me in.

Whatever. I was so bad at burro racing, I wasn't even good enough to be worst. There's a special award for "Last Ass Over the Pass," but you've actually got to get to the pass to qualify. At least I still qualify for the only prize I really want, I thought, as I dragged Blue Note along like a mom with a spoiled toddler: All I had to do was cross that stinking finish line and I never, ever had to see a burro again.

6

The Beastmaster

"*That's* your idea?" Tanya snorted. "A burro race?"

She began squinting as I told this story, squeezing her eyes as if trying not to look at me. "So how far would he have to run?" she asked.

"The World Championship has two distances—"

"The World Championship." She was smirking now, as if she'd just caught the punch line. "Not just a race. A World Championship."

I actually had a good explanation for why the toughest race might be our best shot, but this was already going worse than I'd expected and I hadn't even told Tanya the bad news yet: Secretly, I'd already jumped the gun and taken Sherman on a test run. I hadn't planned to; I wasn't even sure Sherman would be alive when I got up this morning. But when I saw him marching along all morning in low-speed pursuit of Lawrence, I couldn't resist. I got a bag of horse treats and a six-foot lead rope from the feed store. Then I pushed Lawrence out of the way, closing him off behind the gate, and snapped the rope on Sherman's halter.

I grabbed one of the apple-flavored pellets and held it under

Sherman's nose. He sniffed it suspiciously, then snuffled it up and munched, one grinding chew at a time. I took two steps back and offered him another. "C'mere, Sherm," I urged, waving the treat. Sherman looked at me but didn't move. He just stared at me, a battered stuffed animal with sad Eeyore eyes, the Sherman of yesterday again.

I felt horrible. "You're such a dick," I told myself. This poor guy has been suffering his entire life, and now, right when you're supposed to be restoring his trust, you pull a bait-and-snatch? "Sorry, Sherman," I said. I decided to quit messing around and wait for Tanya to get here before I did any serious harm — and that's when Sherman plopped two steps forward and grabbed the treat from my hand.

Game on! I retreated a little farther and dug into my pocket for another treat. Before I pulled it out, Sherman was on me. I fed it to him and backed up again. Treat by treat, he trudged and munched his way across the small grassy pasture until we got to the gate. I swung it open, stepped out onto the gravel driveway, and reached in my pocket. Sherman stretched his neck toward the treat but didn't move his feet. I took a half-step toward him and waved it again.

"Here we go, buddy," I coaxed. I gave the lead rope a gentle pull, but it could have been tied to a tree. Sherman dropped his head, staring at the ground. Something about him suddenly seemed different. His feet were planted, his ears were back; he looked braced, as if digging in for a fight. Whatever bonding moment we'd had during our little snack-and-step exercise wasn't just over; clearly, it was now being held against me as evidence of treachery by a terrified donkey who felt bitterly deceived.

Man, is he scared, I thought. "Okay, free to go," I said. I fed Sherman the last treat and un-clicked the rope from his halter. I reached out to give him a pat, but the second the rope fell free of his neck, Sherman spun away and trotted off, beelining straight back to the back gate where Lawrence was pacing back and forth, waiting for him.

Wow. So he can run. At least a little. And only on grass. I was delighted, but the thrill quickly faded. I had a plan I hoped could bring Sherman back to life, but if his feet were too badly damaged to step on hard surfaces, there was no way I was going to force him. I saw how frightened he was just by looking at gravel, and I wasn't going to put him through that again. Was there any chance he could recover in a few weeks? A few months?

Ever?

Sherman seemed nothing like the donkeys I'd seen in Colorado. To me, they were all tough, mountain-hardened beasts who could run for days and shrug off a tsunami. I couldn't imagine how I'd ever bring Sherman back from his mental and physical trauma—his desperately lonely months locked in a stinking stall and the near-fatal deterioration of his feet—so that he and I, side by side, could race against them in an ultramarathon.

Every time I thought about it, I felt both a thrill of excitement and a knot of dread. Running with Sherman would mean forging a bond with one of the most notoriously cranky creatures on earth, training side by side for big miles on god-awful trails in god-awful weather. I'd already seen firsthand what a mistake can mean when hooves are involved; last spring, our neighbor's son, Elam, was kicked in the face by a mule while adjusting its plow harness. Four surgeries later, his cheekbones are made of plastic and his face remains a sunken mask.

But that's what made it so damn irresistible. My gut told me that the one thing that would save Sherman—the one thing he needed more than petting, or shelter, or even Banamine—was movement. Movement is big medicine; it's the signal to every cell in our bodies that no matter what kind of damage we've suffered, we're ready to rebuild and move away from death and back toward life. Rest too long after an injury and your system powers down, preparing you for a peaceful exit. Fight your way back to your feet, however, and you trigger that magical ON switch that speeds healing hormones to everything you need to get stronger: your bones, brain, organs, ligaments, immune system, even the digestive bacteria in your belly, all get a molecular upgrade from exercise. For that, you can thank your hunter-gatherer ancestors, who evolved to stay alive by staying on the move. Today, movement-as-medicine is a biological truth for survivors of cancer, surgery, strokes, heart attacks, diabetes, brain injuries, depression, you name it. So why wouldn't it also be true for Sherman, with the blood of wild African asses in his veins?

I heard a weird story years ago about Jimmy Stewart, and it popped back into my mind while I was at this crossroads. The story stuck with me because it described the exact superpower I'd always wanted as a kid. When I watched Saturday-morning cartoons and saw Aquaman control fish with his mind, I wished I could do the same thing with animals, summoning all the dogs in my neighborhood to flee their backyards and assemble by my side, a suburban wolfpack awaiting my telepathic command. That was my dream, to become the Beastmaster—and according to Hollywood legend, Jimmy Stewart really was. People said there was something about him that inspired such trust that all he had to do was talk and even animals would obey. Jimmy Stewart had the gift.

It's a crazy story, but if you believe Jimmy Stewart, it's absolutely true. The way he told it, there was a stunt horse named Pie that was so dangerous, none of the other Western movie stars could handle him. "He was a sort of a maverick. He hurt a couple of people," Stewart said. "He nearly killed Glenn Ford, ran right into a tree." But for some reason, Stewart and Pie took a shine to each other. "It was almost a human thing between us. I think we liked each other. I talked to this horse. I know he understood me. I know. I *know*."

What made him so sure? Because of strange episodes like this: Once, Stewart and Pie had to shoot a tricky scene involving a little bell and a band of desperadoes. Stewart was supposed to outwit the outlaws by sticking a bell on his saddle, then slipping off and letting his horse walk into town by itself while he snuck around the back and got the drop on the bad guys. The problem was, how do you explain all this to Pie? Stewart was an actor, not an animal trainer, and there wasn't one on set.

"Well, let me talk to him," Stewart told the director. To Pie, he said, "Now, this is tough because you're a horse, you see, but you have to walk straight down there and no one's gonna be on you. You have to walk right straight down and clear to the other end of the set." The film crew braced for a long night of muffed takes. Instead, they nailed it in one shot. "Pie did it the first time," Stewart exulted. "It was amazing."

Maybe Stewart got lucky. Or maybe he was a world-class bullshitter who knew how to spin a good yarn. But isn't it just as likely, as I tend to believe, that the same empathy and imagination that made him a great actor also allowed him to connect with other creatures? To communicate with them as equals? Toward the end of his career, Stewart made Johnny Carson tear up on *The Tonight Show* by unfolding a sheet of paper from his pocket and reading a poem he carried around about his dead dog, Beau:

> Sometimes I'd feel him sigh and I think I know the reason why.
> He would wake up at night
> And he would have this fear
> Of the dark, of life, of lots of things,
> And he'd be glad to have me near.

That's right; most of us would give a kidney to understand why our dog won't stop barking at the door, yet let a hound yawn at two in the morning and Jimmy Stewart sees right to the depths of its soul. Odd as these stories sound, there's a case to be made that when it comes to being the Pie Whisperer, science is actually on Stewart's side. "Sound sometimes carries emotions across species," points out Carl Safina, the renowned animal behaviorist. "Our shared capacity to perceive it is part of our deep inheritance. Whether the receiving ears belong to a human, a dog, or a horse, several short upward calls cause increased excitement, long descending calls are calming, and a single short abrupt sound can pause a misbehaving dog or a child with a hand in the cookie jar."

For most of human existence, animal intuition wasn't just common; it was a matter of life or death. Our ancestors thought about animals all the time; from the second their eyes opened till they closed at night and during their dreams in between. They had to understand animals, instantly and intimately, or they'd have been wiped out. The story of Beau and Jimmy, basically, is the story of our survival.

To grasp that, consider this: did we train the first dogs, Carl Safina asks, or did they train us? There was a time when humans ranked pretty low as hunters. We were worse than Neanderthals, who were bigger, stronger, and possibly smarter, and wolves, which had superior fangs, speed, and tracking ability. In a Stone Age battle for food, you really don't want to be third on the power chart. But we did have one thing on our side: we were terrific thieves. If we saw a good idea (or a hunk of bison), we stole it. Our ancestors learned to lope along behind the wolf packs, grabbing up the leftovers once the wolves had brought down their prey and gobbled their fill. Over time, we adopted the same harass-and-swarm tactics, but what really vaulted us to the top of the food chain was when we stopped copying wolves and began cooperating with them.

We can thank the wolves for that, Safina contends, because they probably made the first move. Wolves are super curious, and their ability to scent human anxiety would allow them to sniff out the safest moment to approach us. That was a fateful day: our ancestors, crouching and uncertain, watched this alpha beast edging toward them and made a decision that would change the course of history. Like a Hollywood meet-cute, the wariness on both sides soon melted away because wolves, better than any other creatures, really got us. They quickly figured out how we think because their brains were pre-wired with perception skills similar to ours, something Safina calls "human-like social cognition."

And as a team, we were fantastic; we forged a bond that would allow us to dominate the planet. With canines at our side, we became masters of the universe; these new companions were our night watchmen, our GPS guides, our first-wave assault team. In the battle for existence, they gave us the crucial competitive edge that allowed us to persist while our rivals, the Neanderthals, faded to extinction.

After that, well, hell! There was no stopping us. From then on, we couldn't make enough animal alliances. We persuaded horses and elephants to carry us into battle, and hawks and ferrets to kill rabbits and drop them at our feet. Wildcats became tame and protected our grains from rodents. We learned to saddle reindeer and herd geese and yoke yaks. Animals didn't just feed and protect us; they inspired us. We studied them, drew them, revered them. Religions worshipped animals as gods; moralists looked to them for life lessons; seekers chose them as their spirit guides. We named our tribes and babies after them. Egyptian pharaohs went to the grave with them.

Then we forgot them.

The romance that had burned for ages was snuffed in a heartbeat, historically speaking. We had animals on the brain for more than 300,000 years—and then, out of the blue, Edison and Ford came along and stole our hearts. As soon as we had electric lights and affordable cars, we moved indoors and locked the animals out. We don't need guard dogs or plow horses or fresh kill anymore; we hunt freezer aisles instead of forests, fire up the Prius when we need a ride, rely on screens when we crave companionship. Man's best friend became a major pain, an intruder who craps on our sidewalks, barks all night, and has to be confined to cages in the back bedroom and little prisons in our parks. Coco and Cuddles are now another form of trash; according to the ASPCA, more than four thousand abandoned cats and dogs are destroyed *every day*, just to make space for the six million that are dumped at shelters every year.

"Okay, but that's the price of modern life," you might reply, and no one can say you're wrong (a little cold, maybe, but still . . .). I've butchered my own chickens, milked my own goats, and ridden in Amish buggies, and believe me, when it comes to meals and daily commutes, you need only one go-round with those experiences to appreciate how nice it is to have machines take all the gore, udders, and manure off your hands. Our lives became easier, and in many ways even safer and healthier, when technology allowed us to reduce our one-on-ones with the animal kingdom. It took us until the twentieth century, but finally, we one-upped Mother Nature.

Now comes the reckoning.

7

"Earl's in for murder. Several, in fact."

E. O. Wilson first got a whiff of trouble ahead in 1984.

Wilson is a famous Harvard scientist, but he grew up as a country kid in Alabama who was so wild that he went blind in his right eye when he hurt it while fishing but decided it was too good a day to spend in the hospital and kept casting instead of telling his parents. That accident became his superhero origins story: by losing half his vision, Wilson had to change the way he looked at the world. He struggled to see animals in the woods, but his good eye could zoom in so tightly on insects that he could see the waving of their nearly invisible body hairs. He became a world authority on ants, and it gave him a godlike perspective on nature: he could gaze down from above and watch the way an entire society made decisions that would change it forever. It even changed Wilson; he discovered he had become an expert on unintended consequences. Over and over, he would see an ant colony react to environmental pressure by, say, migrating to search for tastier food or escape a threat, and a few generations later, that same colony had evolved into an entirely new species. You never really know what kind of cliff lies ahead, Wilson realized; those ants did nothing but change their address, and that changed their DNA.

By the 1980s, Wilson was getting worried about his own species. He's a naturalist, and the world around him wasn't natural anymore. It was nuts! Given a choice between freedom and prison, we were happily marching into cells and clanging the door behind us. We were shutting ourselves in boxes—cubicles and cars, thick-windowed homes and soundproofed gyms—and cutting ourselves off from the sights, sounds, and smells of the most important relationship we've ever known. Everything we are, Wilson knew, *we are because of animals*. Our brains, our bodies, our conscious and subconscious minds—they all evolved in response to the creatures around us. It was eat or be eaten, which meant we couldn't forget about them for a second. The human nervous system developed as an early-warning animal detection device, scanning 24/7 for any hint of an approaching heartbeat. Animals were simultaneously our dearest friends and our deadliest enemies, and after 300,000 years, you don't just end a bond like that without paying a price. When we close ourselves off from the natural world, Wilson was convinced, we're messing with forces we don't understand. We're changing our address with no idea where we're going.

Wilson called those forces the *biophilia hypothesis*, which literally means "love of living things" but translates more closely to "Your brain may not remember, but your body will never forget that animals have guarded us since the Stone Age." Why do you think cuddling a cat is so stinkin' irresistible? That's your inner caveman speaking, telling you that as long as that kitty is purring, nothing is trying to kill you. Our prehistoric animal partners were extensions of our own eyes and ears, using their sharp night vision and long-range hearing to alert us to danger. Now, when a tabby curls up in your lap or even when you see a cartoon of Snoopy sleeping on his doghouse roof, you're nudged by a calming ancestral instinct that says *Relax, you're safe for now*. Dogs are even more comforting than your guy, at least at night; in 2018, animal behaviorists at Canisius College researched the sleeping habits of people who shared their beds with a pet and found that out of nearly 1,000 women, the majority had "better, more restful sleep" when they cuddled with a dog rather than with their husbands. And not just because the pups are better about sharing the good pillow and turning off their phones: "Their dogs were less disruptive than their human partners," the study found, "and were associated with stronger feelings of comfort and security."

The FBI figured out how powerful this animal-human connection can be a few years ago, thanks to an ailing agent named Rachel Pierce. Rachel is an FBI psychologist with acute rheumatoid arthritis, and she suffers from flare-ups so crippling and excruciating that some days she can't even get to her feet. She wondered if a service dog could help, so she went to a local shelter and found Dolce, a muttley mix of Husky and German shepherd. Rachel and Dolce began training together, and Dolce turned out to be a star: He learned how to turn on the lights for Rachel, fetch her a bottle of water from the fridge, even load her heavy laundry into the washer so she only had to press the buttons. Dolce was such a treasure that Rachel was tempted to break the very first rule she'd been taught about service dogs: *Absolutely No Sharing*. If friends and strangers start playing with your partner and feeding him treats, they can confuse his focus and unravel his training.

Still, Rachel couldn't help thinking about how great Dolce would be with the witnesses she worked with. She often met with children who were recovering from abuse and teachers who'd survived school shootings. Their recollections were crucial, but it's hard to think clearly when you're still scared to death. What if they had a strong, gentle protector by their side? Rachel began experimenting on her own, volunteering with Dolce at nursing homes and camps for grieving children. The Dolce Effect was amazing; as a psychologist, Rachel was thrilled to discover that her dog could create an atmosphere of trust and security much faster than she could on her own. The FBI authorized her to take Dolce along to work cases, like kidnappings and murders, and they became such an outstanding team that in 2012 Rachel received the FBI Director's Award for Excellence.

One thing you notice when FBI agents show up, by the way:

these guys ain't playing. I've reported on Mafia car bombings and, once, the weird case of a Mexican drug kingpin who flew to Philadelphia to have his fingerprints surgically rotated by a local doctor. As soon as the blue raid jackets appear, a chill comes over the crime scene. You can chat with local cops at the yellow tape, but when the feds arrive, you shut up and step back. Every second counts when suspects are on the loose, so FBI investigators are trained to be stone-faced and machine-efficient. Not for an instant will they tolerate some dog underfoot, unless that dog is producing some serious results. And Dolce, they found, built better cases. He took bad guys off the street. Investigators watched as terrified eyewitnesses sank their hands into dog fur and suddenly had better recall and provided actionable tips. With Dolce by their side, children gave stronger testimony in court.

Dolce turned out to be such a star that the FBI added a Crisis Response Canine program to its Rapid Deployment Teams. When a pair of terrorists opened fire at a Christmas party in San Bernardino, murdering fourteen people and wounding dozens more before escaping in an SUV, the FBI raced to address what could have been a national emergency. Two of the first responders were a pair of Labs, Wally and Giovanni. The agents were glad to see them, and not strictly for professional reasons. "People at the command post were working long hours and were under a lot of pressure," FBI Associate Deputy Director David Bowdich remarked to the press afterward. "As the dogs roamed the area, I saw agents and task force members take time out to pet them." Another FBI official was so thrilled with their performance that she had to abandon law enforcement lingo and reach for sorcery to describe their effect: "The dogs have worked a certain type of magic with people under a great deal of stress."

The magic works for the bad guys too. That's the beauty of E. O. Wilson's biophilia hypothesis: The human-animal bond is instinctive in all of us, even if we don't seek it. Or deserve it. Guards at

Lima State Hospital for the Criminally Insane discovered this by accident in 1975 when the inmates in one ward began acting fishy. Lima is for the hard cases, the high-risk deviants who are too violent for the general prison population. "If you're in Lima," as one case manager put it, "you're in for something pretty heavy." But for days on end, one ward was mysteriously calm. Criminally insane convicts occupying an entire ward don't spontaneously decide to simmer down and cooperate, and when they do, the guards get ready for trouble.

Instead, they found a sparrow.

The bird had flown into the ward and injured itself. An inmate snuck it into his room and began nursing it back to health, and it soon became the ward's mascot. Administrators knew they had to confiscate the bird—what if the inmates tortured it, or began fighting over who got to feed it next?—but extracting a pet from a doting gang of arch-criminals was going to be tricky. It wasn't worth sparking a riot, so the guards opted to hold off a little and see what happened. While they waited, the ward remained weirdly tranquil. The hospital decided to push the accidental experiment a little further. It began taking in unwanted animals, accepting just about anything that was dropped off at its doors. Within a few years, Lima was bursting with nearly two hundred winged and furry friends, including goats, chickens, guinea pigs, ducks, tropical fish, two deer, and a goose that had lost a wing to a dog attack.

"Look over there," a case manager said, pointing a *New York Times* reporter toward a prisoner petting a goat named Friendly. "Earl's in for murder. Several, in fact. Then he escaped from a prison and kidnapped a guard in the process." But after the critters showed up, Earl became a new man. Fights and suicide attempts in the hospital had plunged so dramatically that medications were cut by half. Take a second to consider that; it's such a humble solution that you can scoot right past the fact that it's almost Nobel-worthy. Somewhere in the hinterlands of Ohio, a warden up to his eyeballs in assaults and escape attempts decides to take a gamble and introduce some unwanted barnyard animals to the most dangerous men on earth. That gamble could have blown up in his face, but instead of hurting the animals—or one another—the most dangerous men on earth became half as dangerous. Guards and civilians are now safer because Earl isn't holding shanks to their throats, and Earl doesn't need to be pharmaceutically restrained into a drug-induced stupor anymore. Lives are saved and healing is accelerated, all because of that medical miracle worker, Friendly the Goat.

The warden's courage is even more remarkable because at the time, modern psychology considered animal therapy to be a joke. In the 1970s, a psychologist named Boris Levinson of Yeshiva University was trying to convince his colleagues that his recent breakthroughs were mostly due to his dog, Jingles. Levinson had been working with a "very disturbed child," as he put it, when Jingles sauntered into the room. Levinson was struggling to get this child to speak, so he didn't want a distraction. But with Jingles around, the conversation started to flow. Levinson tried Jingles with his other problem patients and got the same reaction: the dog consistently put children and adults at ease and prompted them to open up.

Just like Freud! Levinson remembered a footnote from the history of psychoanalysis: Sigmund Freud didn't consider emotional catharsis a party either. He found it so stressful to listen to his patients' buried traumas that he would sometimes have his dog, Jofi, sit at his feet to take the edge off. After a while, it dawned on Freud that his patients were also enjoying Jofi's company; whenever the Chow Chow was around, even the most tight-lipped were more willing to explore painful subjects.

So with both Jingles and Freud's ghost on his side, Boris Levinson believed he was onto something. He gathered his session notes and wrote them up in a paper for the American Psychology Association. But when he presented the paper at a conference, he was mocked. "What percentage of your therapy fees do you pay to the dog?" someone yelled from the audience. Heckling isn't usually a big part of gatherings like these. Psychology is all about nurturing and bold ideas, so when professionals who tolerate diseased minds all day laugh you out the door, you know you've broken new ground. "The tone in the room," remarked a therapist who was there, "did not do credit to the psychological profession."

Luckily, prison wardens are more worried about shankings in the shower than squabbles among eggheads. As far as they were concerned, if goats and chickens were keeping Lima's prisoners from going over the fence, then bring on the livestock. Penitentiaries soon began following Lima's lead and adopting their own biophilia initiatives. Inmates were recruited to socialize shelter dogs for adoption and train assistance animals for the disabled. Out West, cowboys were teaching prisoners how to gentle wild mustangs for use as riding mounts. After many of these prisoners were freed, they did something extraordinary: they didn't come back. Normally, up to 75 percent of all prisoners who are released will be arrested again within five years. But among prisoners who've worked with animals, the recidivism rate tends to be as low as 10 percent.

Okay, let's tap the brakes for a second. Who says the success of these programs depends on the pups and ponies? Maybe the prisoners just needed an amusing new hobby to pull themselves together, or more open-air time, or a challenging chore to boost their self-esteem. Maybe the animals were just accessories along for the ride. Maybe—and that's why, in the 1980s, two scientists teamed up to find out what was really going on. Alan Beck, a Purdue University psychologist, and Aaron Katcher, a psychiatrist at the University of Pennsylvania, put scores of volunteers through all kinds of ingenious experiments. To test the "open-air" theory, for instance, they divided a bunch of ADHD students with poor impulse control into two groups: one went canoeing and rock climbing, while the other cared for companion animals. Midway through the study, the two groups swapped.

The result: "Contact with companion animals resulted in greater improvement of ADHD symptoms, better learning, and superior school performance when compared with the outdoor experience," the scientists found. But that was just the beginning. "The companion animal experience also provided greater speech gains, better nonverbal behavior, improved attentiveness, and an increased ability to control impulse behavior." *Moreover, these differences were still evident six months after the program ended.* The italics are mine, because holy moly, that's some long-lasting wonder drug.

Yes, "drug." Because here's the wildest part: Beck and Katcher were also monitoring their subjects' physiological responses, and they found that the dogs weren't just entertaining the kids; they were creating a pharmacological reaction. Pet a dog for as few as five minutes and your heart rate and breathing will slow; your blood pressure will drop; your muscles will relax and your breathing will moderate. That's amazingly fast. For a narcotic to relieve stress in five minutes, you'd need a face mask and a tank of nitrous oxide.

Or a jolt of the "love hormone"—oxytocin. That's exactly what was happening, as later studies demonstrated: By petting those dogs, the children were receiving a surge of oxytocin, a hormone that triggers acute feelings of trust, compassion, and affection. Oxytocin is the brain chemical that makes us feel safe and loved. It relieves pain, helps you sleep, and even boosts your immune system, so you're less likely to get sick. Oxytocin is the reason you feel stronger after a warm hug, and why new mothers are able to nurse; that hormonal boost not only helps a mother bond to her newborn but provides a sense of security that it's safe to feed. During those sunny moments when you feel the world is a wonderful place and you're lucky to be here, more than likely you're experiencing a burst of oxytocin.

The love hormone is so powerful that it's even effective against one of the toughest challenges in modern mental health: treatment of combat veterans suffering from post-traumatic stress disorder (PTSD). Soldiers conditioned to "man up" are reluctant to get help, and even when they do, it's not easy to isolate and treat symptoms that can be all over the place, ranging from anger to depression, paranoia to loneliness, acute fear to dangerous bravado. For one out of every three PTSD sufferers, conventional treatments don't help. Luckily, one drug is effective, has no dangerous side effects, and looks rad wearing a bandanna: Jingles. Dogs were actually Plan B; researchers originally thought they could use a nasal spray to shoot oxytocin right up a patient's nostrils. Why have a dog on the couch when you can have an inhaler in your pocket? But whenever we try to outsmart Mother Nature, Mother Nature breaks out the nunchuks. The nasal spray was fine when the dosages were dialed in just right (a group of eighty traumatized male and female police officers in Holland responded beautifully), but even a pinpoint miscalculation turned out to be, literally, a nightmare, causing horrific dreams, insomnia, agitation, and flashbacks.

You don't have those problems with Lugnut, for instance, a three-year-old Golden Lab who has been deployed to help the emotional adjustment of soldiers returning from combat duty. The U.S. military has made such a shift in favor of animal therapies that it has appointed a major to serve as its ranking "Human-Animal Bond Advisor" and established nationwide programs, including Paws for Purple Hearts and Warrior Canine Connection, to work with struggling veterans. Read the testimonials from grateful families and you're guaranteed to wipe your eyes: thanks to these companion dogs, men and women who were too gripped by dark spells and terror to get into a car or go to the mall with their families are now functioning again—and doing so without the dosage mishaps that came from nasal sprays. Outside the military, dogs are also helping sexual assault survivors deal with their own natural anxieties about safety and human contact. Service dogs will rouse their owners out of nightmares and are terrific at providing survivors with a sense of security when they have to turn their backs to operate an ATM.

But where the story takes a peculiar turn is when you look at raw body rebuilding. Take heart attack patients: If you own a dog, you're *twice as likely* to survive the first year after a major coronary incident. What other treatment doubles your chances of survival without a prescription? Cancer patients undergoing chemotherapy need only one hour of animal-assisted therapy a week to see their depression and anxiety reduced by half. Elder-care facilities don't even have to bother with dogs; just put a fish in great-granddad's room, they've found, and he'll eat better, put on healthy weight, and be more social.

Why? No one really knows, exactly. But why do we have to? That's the question that makes E. O. Wilson want to bang his head against Harvard's brick walls. The real question isn't what we get from animals. It's what we lose without them. If the animal-human bond improves our lives in every way—if the sick get stronger, the traumatized feel safer, our children learn faster, our prisons become safer—then the reverse is also true: *Without* animals, we're weaker. We're sicker. We're angrier, more violent, more afraid. We've taken ourselves back in time, back to those desperate days when humans were alone on the planet, peering at wolves and hawks and wildcats from a distance and wishing we could somehow connect. Once we became allies, we had it good—until we turned our backs on the best friendship we ever had.

The good news is, we're trying to fix what we ruined.

The bad: we may have waited a little too long.

Our animal intuition is dead. If you want proof, take a look at Cesar Millan's bank account. The "Dog Whisperer" crawled through a storm pipe from Mexico to arrive here as a penniless teenager, and since then he's donated $1 million to Yale University, mostly to finance a shelter-dog program but partly to show he's now rich enough to give money to a school that already has $27 billion. Cesar made a crazy amount of cash from the simplest job on earth: teaching people how to walk their dogs. More Americans have pets than children; we've got nearly 200 million dogs and cats, and we spend nearly $70 *billion* on them every year, yet Cesar is a living testament to the fact that we have no idea what the hell we're doing. Cesar charges $1,000 a day to teach you how to prevent the dog you brought into your own home from crapping all over it, and his waiting list for seminars is forever. Deep inside, we still feel that ancient yearning to connect with other creatures. But when we try, it's a disaster.

"The dog is not thinking, 'Thank God Cesar finally got here!'"
Cesar explained to me.

"People think I have a special power," Millan has said. "The dog changes when I come in, and for them that is magic. But the dog is not thinking, 'Thank God Cesar finally got here!' The dog is responding to my energy. In the animal world, everything is energy." Cesar himself is a little baffled about why this energy is such a mystery; the only training he ever had came from watching his old *abuelo* handle the hounds on the family farm. Yet look who comes to Cesar for help: Oprah Winfrey, Tony Robbins, Deepak Chopra, Jerry Seinfeld, even John Grogan of *Marley and Me*—all natural charmers who command the attention of millions, but not the hearts and minds of their own pups. Tony Robbins! The guy was life strategist for Mother Teresa and Nelson Mandela, but no match for a four-pound terrier named Missy-O.

And seriously, dogs are a joke. When it comes to training animals, dogs are beginner level, pure bunny slope. We invented them, after all. Dogs are the only man-made species; when we began partnering with wolves, we took over their bloodlines and began tinkering. Whenever we had a new need, we created a new breed; we played Dr. Frankenstein and match-made males and females to create whatever oddball mutations we were hankering for: short legs, snouty noses, fiery tempers, cuddly fur. Meanwhile, business for Stone Age Cesar Millan was already booming; in Saudi Arabia, an early sandstone etching shows a hunter led by his thirteen hounds, two of them tied to his waist, all of them looking well trained and remarkably similar to the Canaan dogs still prized by desert nomads today. Since the dawn of human history, we've been selecting and breeding the dogs we liked best, genetically engineering them to do one thing and one thing alone: obey. After 30,000 or so years, we've created the MacBook of the animal kingdom, a creature ready to operate right out of the box.

Dogs, in other words, ain't no donkeys.

We've been messing with donkeys for only a few thousand years, just a fraction of the time we've invested in dogs, and the results have been impressive: the donkeys have barely changed a bit. You could stick Sherman in a field full of his African wild-ass ancestors, and in about five minutes you wouldn't be able to pick him out from the crowd. Donkeys are so tough to tame, we saved them for last; first, we toughened ourselves up by domesticating just about everything else with hooves—cattle, sheep, goats, even llamas. When North African tribes finally gave it a try, they realized donkeys have two very special qualities:

- Wow, these things are *strong*.*
- And freaking deaf.

Donkeys don't react; they reason. They're not like horses, which can be forced to obey out of fear. They're not really stubborn; they're survivors. If a donkey smells danger, its first and fiercest

* Loosely translated. Not to mention speculative.

instinct is to turn to stone. You can force a horse to leap into a river, but if a donkey doesn't know exactly where it's going to land, its feet won't leave the ground. For wild donkeys on African plains and mountain ranges, that insta-freeze instinct was a terrific survival adaptation; predators couldn't scare them into plunging off a cliff or revealing their whereabouts, since the less they moved, the better their dun coats blended into the landscape. Horses are speedy, but when it comes to steadiness, stamina, and heroic resistance to heat, cold, and thirst, you can't do better than a burro. Nothing we've done to domesticate donkeys has weakened that instinct, which is great: it makes them idiot proof. No wonder misfits and monarchs of all stripes—prophets and prospectors, conquistadors and trappers, hermits and explorers, Jesus and his mom, King Solomon, the prophet Muhammad, and even Queen Victoria—made long-ears their transport of choice.

Even George Washington, "the greatest horseman of his age," according to Thomas Jefferson, was privately a donkey guy. Away from the battlefield and back home on the farm, our first president was also our first donkey aficionado. King George of Spain gifted a pair of burros to Washington, who liked them so much that he built them into America's only breeding herd. Washington was outmatched, though, by an ancient Egyptian king whose burial place was discovered not long ago. When archaeologists opened the tomb, they expected to find the king's favorite courtiers buried in the "high-status area" around him in a protective ring. Instead, he was surrounded by ten beloved donkeys, sacrificed to be his guardians in the afterlife. Do you know what other animals were ever given that honor? None.

Ask Curtis Imrie on the wrong day, on the other hand, and you'll hear a very different theory about how those donkeys met their end: "If you want to explore your capacity for murder," he says, "try a burro race." No one knows more about the donkey mind than Curtis, who bred his own champions from wild stock and ran the twenty-nine-mile Fairplay race for an astonishing forty-two consecutive years. After four decades, though, Curtis would grade his own mastery at just a notch above okay. Once, Curtis was nearing the finish of a fifteen-mile race in Buena Vista, Colorado, when his burro, Jackson, suddenly skidded to a stop in front of a wooden bridge. Nothing Curtis did could persuade Jackson to put one hoof on the bridge, even though it was an out-and-back race and Jackson had already crossed the same bridge, no problem, less than an hour ago. Curtis finally had to tie Jackson to a tree, walk into town, and return with a winch-equipped Jeep to pull the 750-pound animal across the bridge, one slow crank at a time.

"Indians didn't have donkeys," Curtis says. "They saw ours and thought they were called 'Goddamnyous.' Donkeys know their rights and they can shut you down fast." If a pro like Curtis got stonewalled by Jackson, what hope did a rookie like me have with Sherman? But imagine the payoff if I could pull it off. If I could crack the donkey language barrier and get Sherman to join me for the adventure, I could show the way for everyone else who wants to partner with an animal but doesn't know how to begin. Whatever that Stone Age hunter did to inspire his loyal pack, whatever secret Cesar Millan whispered in Tony Robbins's ear to sync his brain with Missy-O's, that's the kind of knowledge I could learn by bonding with Sherman.

There was no way I could do it on my own. I had to find a Donkey Whisperer, and I knew only two. One was Curtis, who was 70 years old and 2,700 miles away. The other was Tanya, who was right in front of me but staring daggers. If I couldn't change her mind, this project was dead before it started. How would I get Sherman to race through the Rockies when he wouldn't even step onto the driveway?

8

The Barely-a-Puddle of Doom

"That might be his brain, not his feet," Tanya said.

Four days had passed since Sherman arrived, and three since I'd nearly made Tanya's brain explode by telling her that I'd tried hauling her ailing patient with the crippled hooves into the road to see if he felt like doing some jogging. I could tell she still had some doubts about me when she came by that morning to examine Sherman's hooves, but after she'd lifted each one in turn to check the condition of the soft flesh and the shape of the trimmed hoof, she stood up with a smile.

"Looking good," she said. "Scott worked a little miracle."

"Soooooo—" I ventured. "You think there'll come a point when he'll be able to walk on the road? Maybe even run a little?"

Tanya wasn't on board yet with my burro race idea, not by a long shot, but I could tell she was intrigued by the intellectual challenge. For a skilled trainer like her, it was like tackling a math problem for NASA; she wasn't promising she could put a man on Mars, but she wanted to at least see if she could crack the equation.

"You've got no idea what kind of life he's had," she responded.

"There could be a lot of trauma between those ears that needs to be unpacked." She glanced at the driveway, then back at Sherman. "Okay, let's try something. Go get that goat he likes."

I fetched Lawrence and snapped a rope on his collar. Lawrence is always up for an adventure, especially when it might involve a treat, so he trotted eagerly beside me. We passed right by Sherman, but oddly, he didn't even give us a look.

"Bring him out here to the road," Tanya said, "and let's see if Mr. Sherman follows."

I could tell Sherman was spying on us, because as soon as Lawrence and I got a few yards away, he began ambling nonchalantly in our direction. Sherman picked up the pace as Lawrence and I got closer to the gate, first shifting into a full walk . . . then an anxious march . . . and then he charged, coming at us like a bull. I froze, bracing myself for a butting, then realized he was heading for the gate. *Jesus, is he escaping?* Sherman brushed past me, then skidded to a stop and turned sideways, blocking the gate with his body. His head dropped back down, his ears drooped, and just like that, he turned back into sad, innocent Eeyore.

"Oh my god!" Tanya shouted. "I *love* it. Look how he's playing you!"

Lawrence still had eyes on Tanya, convinced that she was packing treats, but when he tried to get to her by squirming past Sherman and passing through the gate, Sherman wedged himself into the gap and refused to move. Sherman wasn't letting Lawrence go anywhere without him, which meant that if Lawrence didn't understand how dangerous the world was outside that gate, then Sherman's only option would be to spring into action and save them both.

"Shermie, you rock star!" Tanya laughed. "All messed up but wicked smart. Dude, he's like the Good Will Hunting of donkeys."

The wheels in her mind were turning. She gazed at Sherman, processing that little stunt he'd just pulled. A lot had been stolen from Sherman during his years in captivity. His muscles were withered, his body was sagging and soft, his trust was severely damaged if not altogether lost. But that? That mad dash to rescue Lawrence? Whatever else he'd lost, Sherman seemed to be showing that he was still determined, brave, and amazingly loyal. Not to mention kinda crafty.

"So when is this race?" Tanya asked.

"Next July," I said. "Little less than a year."

Tanya pursed her lips, rocking her head back and forth noncommittally. "How far will he have to run?"

"Twenty-nine miles," I said. "Fifteen for the short course."

"Fifteen. Is short." Tanya rolled her eyes. "Well, I'm the one who told you to find him a job. But it won't be easy. Sherman can come up with a million ways to make your life a living hell. See what he did when you tried to get him out the gate? He's already two jumps ahead, and you haven't even started."

Tanya was back in man-on-Mars mode. "There's only one way this even has a chance," she said. "Anything you want a donkey to do, you've got to make him think it's his idea. Let's try something."

She walked back to her truck and loaded up a fanny pack with horse treats. With one hand, she fed a few to Lawrence to get him out of the way, and with the other, she gave Sherman's ears a good, friendly scratching. After a few minutes, Sherman seemed to relax a little. Tanya fed him a handful of treats, then unclicked the lead rope from Lawrence's collar and snapped it onto Sherman's halter. She stroked his face one more time, and then she stood tall with authority to show him who was boss and led him . . .

Nowhere.

Tanya pulled on the rope, gentle but insistent. Sherman dropped his butt and locked his front legs, hunkering down for battle. "That's fine," Tanya said. "Now we wait." She held the rope firmly but didn't tug, parrying Sherman in an evenly matched tug-of-war. "Wait for it . . ."

Gradually, Sherman relented. He took one step, then another, until he reached Tanya at the end of the rope. "Good man!" Tanya crooned, feeding him a handful of treats. When she began walking again, Sherman followed right behind her.

But when we reached the road, Sherman recoiled like it was a river of lava. "Maybe he's never seen asphalt before, maybe he thinks it's a bottomless lake, who knows?" Tanya said. "But with a donkey, anything you start you have to finish." She stepped into the road, looped the rope under her butt, and sat back on it while Sherman threw every fiber of his being into reverse. One of my neighbors rumbled past on his tractor, and he had to yank the steering wheel and swerve wide to avoid running Tanya down. Tanya smiled and waved, holding her ground in the middle of the road.

Finally, Sherman placed one foot on the asphalt.

"Sherminator!" Tanya crowed. "Good man. See? You trusted me and you didn't die."

Tanya walked toward Sherman, reeling in the rope as she approached so he had to keep that one tentative foot planted on the road. "Right now, we've got a choice," Tanya told me. "It's the biggest choice you'll ever make with Sherman. Whichever way you go, it's going to determine the relationship you have with him for the rest of your lives." This was the time to decide, she said, between "easing" and "flooding."

Tanya held Sherman in the same spot, with that one hoof on the pavement, while she explained the difference to me. "Flooding" an animal means bombarding it with new sensations, forcing it to follow your commands without giving it time to process what's happening. If your dog gets feisty and nippy around loud voices, for instance, then it's time to break out the pots and pans. You're supposed to hold your pup tight on a short leash and subject it to such a terrifying barrage of clanging that in the future, every other noise will seem trivial by comparison. Flood the dog's sensory system with that decibel overload, and its fear of noise should be gone forever.

For Sherman, the flooding approach would mean goading him to keep moving forward across that blacktop road no matter what strange dangers lay ahead: the red Stop sign, our neighbor's gigantic plow horses, the rumbling creek across the street, the creaking metal road sign for AK's Saw Shop. Sherman would have alarm bells going off like crazy in his mind, but too bad: he'd learn that when I talk, he walks. Period.

"Easing," on the other hand, would reduce that flood to a trickle. It's better known as "whispering," but these days there are so many self-appointed whisperers out there beating their chests about their magical powers with dogs, cats, macaws, CEOs, and psychopaths (yes, there is a "Psychopath Whisperer") that no one can agree anymore on what qualifies as whispering or who's qualified to do the whispering. Tanya isn't a fan of flowery phrases anyway. She'd rather stick with straight talk, and to her, "whispering" has nothing to do with crooning in the animal's ear and everything to do with taking things slow and easy.

Easing would give Sherman a chance to process the world at his own pace. Instead of being driven, he'd be guided. We wouldn't move on to the next challenge until he'd sniffed the last one, and eyeballed it for a while, and finally signed off. Easing is a more gentle approach, but it has a downside: it would force Sherman to face his fears head-on. He couldn't just shut off his brain and follow blindly, a servant to a dominant master. He'd have to be fully aware of everything we were doing and make his decisions on the basis of courage, not intimidation. The choice isn't as obvious as you might think, especially for a mistreated animal like Sherman. For him, letting his mind go blank could be a relief.

I'd seen the Temple Grandin biopic, so I had some sense of what Tanya was saying. Glancing around our home, I tried to see it through Sherman's eyes, scanning the landscape for little triggers that wouldn't register on me but would freak out a more hyperaware, born-to-flee prey animal. About a hundred yards from our driveway was a dirt road cutting through the woods. It would be the perfect place to take Sherman for walks, except if we went full-on whisper, it could take us a week just to get him there. Every few steps along the way there was something new that he would have to absorb, and I wasn't psyched to stand by the side of the asphalt road, dodging pickup trucks, while Sherman weighed the risks of AK's Saw Shop sign.

Easing would put us on a very *looooong* calendar, and we didn't have time to spare. Flooding would be a lot snappier; the only limit to Sherman's training would be our own capacity for relentless persistence. And the time to start was right now, immediately, while we had that first hoof on the pavement.

Or not.

"Okay, Shermie, go chase a goat," Tanya said. "Time to call it a day." She turned Sherman around and led him back to the gate. Lawrence was still waiting, his hooves on the rails and his head poking over the top. Apparently I'd misread Tanya completely. For her, all that stuff about flooding versus whispering was an academic exercise, not a real question. There was never any doubt in her mind about what Sherman needed.

Easing was the only way to go, Tanya told me, as she opened the gate and turned Sherman loose with Lawrence. "I'm not going to brutalize something just for my entertainment. Sherman would never go for it anyway. He's had enough tough times in his life." Tanya promised she'd be back in a few days. Until then, my homework was to build on what we'd accomplished and see if I could defy all odds and, whisper-style, get Sherman to place all four feet on the road.

"This is really important," Tanya reminded me before she headed off. "Anything you start, you have to finish. If you don't think you can finish a task, don't start it. Got it?"

"Sure," I promised, not especially concerned. What was so hard about sitting on a rope? To be honest, for all Tanya's *Sturm und Drang* about easing and whispering and flooding, all she really did was lean back and give Sherman cookies. "Sitting and feeding, that you can handle," I told myself, never realizing that like Prometheus, those whom the gods would destroy, they first make sassy.

For the rest of the week, the main task I had to finish was keeping Sherman's butt under close surveillance. I recognized right away that only David Blaine could pull off Tanya's sleight of hand with a rectal thermometer, but fortunately she didn't ask me to try. As long as Sherman was eating and moving around, she said, we could hold off on checking his temperature until her next visit. In the meantime, my job was to remain on Poop Patrol. I was supposed to keep an eye on Sherman's stools and classify them as one of three categories: Normal, Weird, and None. Weird was anything runny, yellowish, or downright messy that could indicate he needed another dose of dewormer. None was a red alert; that meant his intestines were impacted and I needed to call Tanya ASAP.

"And Normal?" I asked.

"Is normal." Tanya shrugged. "A good, healthy plop."

"I've never seen a healthy plop," I pointed out. "I've only seen Sherman."

"You'll know it," she said. "You'll look down and go, Yup, there's a good one."

I did my best to follow Tanya's instructions, but as far as I could tell, nothing that came out of Sherman for the next few days seemed worthy of panic or applause. That was something, at least. Sherman was still shy around the sheep and steered clear whenever they were marauding at the hay feeder, but now that spring grass was coming in, Buddy the ram and his crew were spending more time ranging across the meadow, leaving Sherman and Lawrence on their own to nibble together in peace. Tanya stopped by now and again, checking his temperature and probing his belly.

"He's on the mend," she pronounced. "Now's the time to get him moving."

So when my daughters got home from school that afternoon, we stuffed our pockets with horse treats and headed for the pasture. We haltered Sherman up, then shoved Lawrence back behind the gate and led Sherman toward the road. My fourteen-year-old daughter, Maya, held the rope while Sophie, the ten-year-old who'd gotten us into this, went in front to check for oncoming cars.

"You've just got to sit back on it and—" I began, reaching for the rope to loop it under my butt and demonstrate Tanya's passive-insistence strategy. While I was talking, Sherman just walked right past me. He put one foot on the road . . . then another . . . and then he was off, accepting Maya's lead and following Sophie into the middle of the road, clumping along like he'd been strolling behind her all his life. Either we were world-class whisperers or something else was going on, which meant something else was definitely going on. We walked a little farther and then stopped to give Sophie a turn leading him. I showed her how to hold the rope: braced across her hips, firm in her hands but loose on his head.

"And then," I said, "you just go."

Except Sherman didn't. I gave the rope a firm pull—nothing. Sophie tried, too, but for both of us, Sherman was a boulder. What the heck was Maya doing that we weren't? Sophie handed the rope back to Maya to see if she could get Sherman going again, and a moment later, the mystery was solved: when Sophie stepped out of the way, Sherman stepped with her. Sophie stopped, and so did Sherm. We switched it around and let Maya take point, and Sherman was just as quick to follow her. Me, he wasn't all that keen on; I tried getting in front and Sherman slowed down so much, he looked like claymation. I can't say I blamed him: by now he knew that whenever I showed up, someone pulled out a hacksaw or took Lawrence away.

Sherman seemed happy sandwiched between the girls, so I moved aside and let them do their thing. Tanya hadn't left me with a playbook for this; neither one of us had anticipated the problem of Sherman suddenly showing a taste for afternoon hikes. I was a little uneasy about how far to let them go. Would his hooves hold up? Or was it smarter to end the fun now and turn the girls toward home? The dirt road was *so close*, though—twenty more steps and we'd be there. It was a little risky; we were heading into a nasty blind curve that pickup trucks liked to take on two wheels, but if we kept moving, we'd be around it in a jif.

"What's wrong?" Maya asked. "He looks so scared."

The moment we hit the curve, Sherman froze, with his ears flattened like a nervous cat's. He was doing this weird hunching thing, pulling his body into a big furry ball but keeping his legs stiff, like he was trying to both disappear and fight to the death at the same

"Easing" Sherman past his fears

time. I glanced around and tried to Temple Grandin things out, searching for anything that could be spooking him. There were no dogs. No tree-caught plastic bags. No—

Wait, not that puddle? That wet stain? It was barely a foot wide. It didn't even have any water. But Sherman was rooted to the spot like a kid peering over the edge of a high dive.

"Maybe we should take him home," Maya said.

Absolutely. Any second now a truck could come blasting around that curve and find two kids and a donkey in its path. Except Tanya's last words were still ringing in my ears, and I didn't want to spoil the amazing leap that Sherman had already made today. "Let's just get him across the stain first," I told the girls. "Tanya showed me how." Two little steps and he'd be through the thing. I slung the rope under my butt, ready to sit back and get this over with.

But I wasn't the only one who'd learned something that morning. When Sherman saw what I was up to, he yanked his head back before I could get set and threw me off balance. I barely caught myself before sprawling on top of him. *Son of a—*

I took a breath and white-knuckled the rope. Okay, now I knew two things for sure: the problem was the stain, and I had to do something about it. Sherman didn't seem scared; he looked defiant, like a kid clamping his lips and deciding he's not taking his medicine, no matter what. His mane was bristling and the look in his eye said that he wasn't giving up so easily this go-round. You better get this under control now, I thought, or you never will.

"Girls, watch the road for cars," I said, then thought better of it. "Actually, get back over here next to the fence." This could take a while, and I didn't want the girls straying anywhere near the blast zone of Sherman's hooves or oncoming speeders. I hunkered down on my end of the rope, pulling it taut until Sherman's head was over the stain, although his feet remained stubbornly at the very edge. For an ailing animal, his strength was unbelievable; I had to dig deep just to make sure I wasn't the one being dragged down the street. We were locked in a grim standoff, staring each other in the face, neither of us gaining or losing an inch. Two minutes ticked by . . . three . . .

My hands began cramping, and a part of me began to wonder what I was doing. This was exactly the kind of situation I'd been desperate to avoid for half my life. Twenty-five years ago, I was skiing with my brother in Jackson Hole when we saw an advisory for backcountry snowmobilers: "Under no circumstances should you leave your vehicle," it warned. "Visitors have been gored to death after approaching buffalo on foot." Somewhere, grieving wives were telling their kids that Daddy wasn't coming home because he tried to pet a wild bull. Ever since then, I've been phobic about dumb-ass deaths. I didn't want Mika to have to stand by a casket and explain to family and friends that no, I'd been in perfect mental health, but yes, my final moments were spent wrestling in a puddle with a donkey.

"Car!" Sophie called. From a distance, I recognized the rattle of our seventy-year-old neighbor's old farm truck. I knew Sam wouldn't be moving fast, so I stayed put. Sam must have been used to public man-animal standoffs, because he came around the bend with a friendly wave and a toot of the horn and kept going.

Sherman jumped at the sound of the horn, and it was just enough to swing the momentum my way. I hauled back before he could steady his feet, hand-over-handing the rope till he was on my side of the stain.

"Treats!" I called to the girls. "Let's feed him and get out of here."

The girls dug in their pockets. Sherman snuffled up the treats while I scratched his head and said, "Good boy," remembering Tanya's command that no matter how much grief Sherman put me through, every session had to end with a smile. We all crooned and petted him, then turned for home. As the girls began walking, Sherman hurried to stay with them, following so closely that his snout was nearly riding on Sophie's shoulder. Sophie walked faster, then a little faster, until all four of us burst into a run.

The girls were laughing, tickled that no matter how fast they went, Sherman's bobbing head was right behind them. We raced toward home, thrilled by a music that none of us had ever heard before, not even Sherman: the drumbeat of his four flying hooves.

9

Donkey Tao

That night, I went to bed with a problem. I woke up with a plan.

Before falling asleep, I'd thought back with satisfaction on everything the girls and I had accomplished that day. In only his ~~first~~ ~~second~~ third attempt at getting on the road, Sherman had conquered his dread of pavement and crossed the Barely-a-Puddle of Doom. Best of all, he'd shocked the three of us by suddenly bustin' loose and running all the way home by our sides. Sure, it was only a few hundred feet, and yes, he was eager to get back to Lawrence, but still, it must also mean he was becoming comfortable with me and the girls and considered the little brown barn his home.

As I dozed off that night, I imagined what it would be like for Sherman and me to run the World Championship together. One by one, I could see each obstacle on the racecourse falling away behind us. I pictured myself soothing Sherman through the shotgun start, then urging him over that first mile of pavement until we reached the trails. We would really hit our stride on the dirt, leading us to the first of several—

My eyes flashed open. *The creeks.* I'd forgotten all about Colorado's gorgeous, cascading, mountain-fed pain-in-the-ass creeks. There were at least two creek crossings in the World Championship, maybe four, and depending on snowmelt and weather on race day, they could be anything from ankle-deep and glassy to bottomless and thundering. Sherman had just fought me like a grizzly over a splotch in the road; how many weeks of pulling on deadweight would it take for me to actually get him into moving water? We didn't have that kind of time to waste; he and I both needed to log a ton of running miles in the next ten months to have any hope of finishing that race, and already the clock was ticking. The damn creeks brought us right back to square one, back to the same whisper vs. flood, drill sergeant vs. den mother predicament we'd had from the start. If I let Sherman ease through his aquaphobia at his own speed, we'd be standing around *for hours* in endless tugs-of-war. But if I forced Sherman to run, would he ever want to?

By the time I fell asleep, I was feeling hopeless again. When I opened my eyes, I had my answer: the Tao of Steve.*

My mind must have been churning all night, because it somehow connected that old memory from fifteen years ago with a little throwaway nugget of advice that Tanya had given me the day before. "You've got to make Sherman believe everything is his idea," she'd said, while I nodded along and yeah, yeah-ed without really listening. You hear that kind of stuff all the time in heist films and rom-coms and World Series of Poker commentaries, but c'mon—who besides George-Clooney-as-Danny-Ocean is really clever enough to trick people into insisting that no matter how much you "object," they're absolutely going to put your briefcase with the secret compartment into their casino safe. My sleeping brain must have cracked the code for me, though, because when I got out of bed I knew exactly what to do.

My mistake was thinking of Tanya's advice as a trick. When I first met Mika and went Tao-style, I genuinely put myself aside.

* Chapter 3. Remember?

Mika loved African music and didn't want some stranger creeping on her, so I loaned her my Bonga and Cesária Évora CDs and made myself scarce. Whatever happened next was completely up to her. If I'd expected anything in return, there's no way the Tao would have worked. I'd have been anxious or resentful or pushy, calculating all my actions against her reactions, radiating a hey-what-about-me vibe that sooner or later would have ruined everything. Done right, on the other hand, the Tao of Steve makes you forget the future and focus on the moment. You're not playing a trick; you're putting your mensch foot forward.

I couldn't wait to hustle outside and get started. It was a beautiful September morning, already getting toasty at nine o'clock. Perfect for my plan. Sherman and I were going to shake the dust off my old playbook and, fingers crossed, repurpose it as a training method we could both get behind: Donkey Tao.

The girls were in school, so I enlisted Mika to help. We grabbed a couple of halter ropes and some horse treats and went to assemble the rest of our team. Lawrence and Sherman were browsing on the far side of the pasture, and as I expected, Lawrence's head shot up as soon as he heard us; he'd figured out that ever since Sherman had arrived, we always showed up packing treats. Lawrence broke off and frisked over at a trot. Sherman followed, and when he got close enough, we slipped the new purple halter I'd gotten him over his head and clicked on a rope. I left Sherman with Mika while I went searching for the next ally I needed for the Donkey Tao experiment: a grumpy white goat prone to fainting spells named Chili Dog.

Chili had joined our family thanks to a pretty crafty prank pulled by an old gent named Ken Brandt. A few years earlier, Ken had bought two of our most adventurous, fence-defying goats, Skeedaddle and Lulu, because he wanted them for his great-grandkids. Exactly *why*, however, was a story of its own.

Ken lived in Falmouth, a village along the Susquehanna River that's even smaller than Lancaster and about an hour deeper into the central Pennsylvania countryside. Back in the 1970s, Ken and his buddies used to tease one of their friends for fishing and hunting all the time instead of taking care of his lawn. One weekend, Ken and the guys pranked their pal by staking a pair of goats on his overgrown front yard. The gag kind of fizzled because the goats were adorable and the guy they'd punked now had free landscaping, so Ken had to up his game. The next time their friend was off in the mountains, Ken put an ad in the local paper announcing the "1st Annual Falmouth Goat Race," with instructions to call his friend's home phone for date and location. This time, Ken struck gold. His buddy returned Sunday night and was blasted by his irate family, who'd spent all weekend answering a barrage of calls about some phony goat race.

Most pranksters understand that when you piss off your victim's family, that's your cue to either back off or play dumb, but most pranksters aren't Ken. Ken didn't settle for snickers; he reached for greatness. His deception had revealed a hidden truth: apparently, a lot of his neighbors had both goats and a competitive streak. If they were motivated enough to pick up the phone, Ken figured, why not give them the race they wanted? He knew the perfect spot too: right down the middle of Falmouth's only road, all the way to Falmouth's only Stop sign. Ken put another ad in the paper, and this time it was his own phone that rang off the hook

"Quite a few people wanted to register," recalled Ken's wife, Jean Brandt. "We really had no idea how many people around here had goats, or how far they were willing to come to race them." At that first race in 1978, Ken didn't fuss around much with rules; he shoved the adults into one race, kids into another, told them to hold tight to their leashes, and pointed to the Stop sign about forty yards away.

"Fine people of Falmouth, are you ready?" he shouted.

"READY!"

"On your marks. . . . Get set. . . . GOAT!"

Nancy Sweigart was standing on the sidelines that day. She'd always been a horse person—and if she had a second favorite animal, it would definitely be dogs—but once you see your daughter's math teacher sprinting her heart out alongside a two-year-old pygmy goat to steal a photo finish from a fifty-three-year-old, four-term state representative who's lunging so hard he nearly eats asphalt and finishes with his chest heaving like a fireplace bellows but still kneels after the loss and immediately, although he can barely breathe, pets his spotted Boer goat and pants, *It's not your fault, Bobo, it's not your fault*, well . . . how can you resist?

"It really took hold of me," Nancy would tell me. Even though she was thirty years old at the time and couldn't remember when she'd last moved faster than a walk, the goats inspired her. She got a pygmy of her own, Bubba, and because nobody really has any idea how you're supposed to teach a goat to race, Nancy invented her own method. "I'd sneak up behind and goose him. He'd take off running and I'd chase him. Then he'd chase me. Then he'd jump on the car and dance around on it till my husband came out and we had to stop."

Lest you look down on butt-thumbing and car-hood rave parties as crude and unscientific fitness strategies, consider that Nancy went on to reign as a three-time Grand Champion who never missed a race for fifteen consecutive years. In all that time she never thought of herself as an athlete, but she was having so much fun tearing around the backyard with first Bubba, then Bear, and then Barney that she was putting more hours and sweat equity into her workouts than someone who spent the morning at spin class. When I met her in 2017 at the thirty-ninth running of the goats, she was nearly seventy and introducing her ten-year-old granddaughter, Autumn, to the sport.

Autumn had borrowed a little white goat named Johnny, but in their first heat, Johnny suddenly stopped after five yards and refused to move. The other children were all flying with their goats to the finish line, while Autumn was stuck in the middle of the racecourse, embarrassed and confused. That's when Nancy showed Autumn what champions are made of. She pushed through the crowd of spectators, ran to her granddaughter's side, and scooped Johnny up. With the little goat in her arms, Nancy jogged with Autumn to the finish line as the crowd roared them home.

"See, everybody loves him!" Nancy told Autumn, who was smiling now. "He's too cute to race. We'll just hug him instead."

Ken Brandt's little gag grew into a monster. By the fifth year, the side streets around Falmouth Road were too jammed with pickup trucks hunting for parking and kids chasing runaway race partners, so Ken moved the event out of the village and off to a converted horse pasture on the outskirts of town. He added a New Year's Eve event, the Dropping of the Goat from a thirty-five-foot flagpole, and knighted himself the first Keeper of the Goat—or Scape Goat—entrusted with safeguarding the ceremonial toy ram under lock and key because, as Ken would say, "New Yorkers in Times Square might get jealous and come to kidnap him."

With extra space, the race got yeasty and swelled into a local Lollapalooza. Hundreds of cars now snake onto the fairground every September, releasing gangs of children who charge off to the petting zoo, the tractor-cart rides, the homemade steam-engine ice-cream maker, and the Tootsie Roll spitting contest. The U.S. Naval Academy mascot, Bill the Goat, has made an appearance, as has a bewildered young couple from Hawaii who wandered in while visiting for a friend's wedding and, as part of Ken's mission to keep things weird, found themselves appointed official finish-line judges for a competition they'd never heard of before.

Ken had one nagging problem: satisfying demand. First-timers would show up just to have a laugh, and by the end of the day they'd be peppering him with questions about how and where and if they could get a goat of their own. All you have to do is feed a fistful of corn to, say, a floppy-eared Nubian, and you'll soon realize that just about everything you ever wanted from a dog, you get from a goat. Goats are affectionate, gentle, and playful. They run and jump and play, but don't bite, bark, or fight. Goats won't bother your cats or attack the mailman, and they'll do you a solid and clean out all the poison ivy and ragweed you've been meaning to pull. And not to beat up on dogs, but when's the last time your Golden Lab filled your fridge? Because if you're willing to get hands-on, you'll find that goats are easier to milk than cows, yet still give up to a gallon a day of sweet, easy-to-digest (good-bye, lactose intolerance!), cheese-ready goodness.

Granted, you could accidentally get yourself a Bamboozle, but if you choose wisely and have the residential zoning, a little cud-chewer could be the pet of your dreams. Here's how loving goats can be: they're even fans of your pee. A few years ago, rangers at Glacier National Park in Montana were mystified when wild mountain goats stopped fleeing tourists and instead clambered down from the high peaks to hang out near them. Ordinarily it's hard for tourists to even spot wild goats, but suddenly they were all over the place. Some tourists were even spooked when they stepped into the woods to relieve themselves and found a gang of goats "lurking" there. What were they up to? Goats generally aren't attracted to human food or close quarters, so they weren't after snacks or shelter. Maybe they saw humans as protection against bears, wolves, and mountain lions? Could be. So Colorado State University sent a scientist in a bear costume* to investigate. That's right: a scientist. Dressed like Yogi. Chasing goats.

The professor growled and ran around and discovered that yes, goats actually do stick close to humans when scared. But things got really interesting during his breaks: when the goats were no longer pursued, they sniffed around until they found a rock or tree with a smelly spot, and then got to licking. Goats, the researchers realized, are attracted to human urine. Our super-salty diets make our whiz rich in sodium and minerals, which the goats crave. "It was as if the sound of my zipper was a dinner bell," reported a hiker who

* Actually, there's a proud tradition of biologists going full method actor in the name of science. Redouan Bshary of Switzerland's University of Neuchatel, for instance, produced groundbreaking research on monkey distress signals by crawling through African jungles in a leopard skin.

was freaked out by a similar experience on Ingalls Peak in Washington's Cascade Range. "I didn't finish, or even start. I zipped my pants back up and briskly walked away." Too bad, because if he'd been a little more cocksure,* he could have witnessed the entire history of animal-human partnerships unfold right before his eyes, a romance built on salt, safety, and chèvre.

You can't really blame the guy for performance anxiety, though: goats have a gaze that's a little too human for comfort. It's not their eyes, exactly; humans have round pupils, well suited for long-distance hunting, while goats have horizontal slits, giving them a panoramic view of approaching threats. No, it's the *way* goats look at us: goats have a rare ability to speak to us with their eyes. Farmers have known this since the dawn of civilization, but only recently have PhDs come up with proof. Goats can communicate with humans in a "referential and intentional way," according to Christian Nawroth, who researches animal cognition at Queen Mary University of London. If you present a goat with an unsolvable problem, it won't just hoist its tail and walk off, the way cats do;† instead, it will look you dead in the eye and wordlessly ask for help. Try it yourself: all you need is a goat and a Tupperware of pasta.

"They go crazy for it," explained Nawroth, who designed the study. "Some goats like apples, some don't. I haven't found any goat that does not like pasta." Nawroth would open the Tupperware, feed the goat some pasta, then snap the lid tight and watch what happened next. Nearly all the goats reacted the same way: they'd look back and forth from Nawroth to the Tupperware, over and over again, rotating their heads in the same "gazing behavior" you'd use to warn the cops that a burglar was hiding in your closet.

Human toddlers communicate this way. So do adult dogs and some horses, but they're a different case: they've been selected

* Yes, intended.

† "Cats performed poorly and barely looked at humans, potentially owing to their rather solitary lifestyle," reported animal cognition researchers.

and bred for centuries to perform complex jobs by our side. With goats, it's natural; they just seem to believe they can reason with us. If they didn't, Ken Brandt's race never would have had a chance. Before the goats are led to the starting line for the first time, very few have ever run alongside a human before. You'd think their first instinct, as prey animals, would be to act like one; you'd expect them to balk, or fight, or flee crazily and tie everyone up in a giant knot of tangled leashes.

But that's not what happens. And that's why the Falmouth Goat Race was such an immediate—and accidental—success. It wasn't thanks to Ken, who never expected more than a laugh. It was thanks to the goats, who figured out the game and agreed to play along.

Ken has given away so many starter goats over the years that he has to keep rebuilding his own breeding stock. That's what brought him to our house. Even though we live more than an hour from Falmouth, Ken spotted our ad on craigslist and decided it was worth the drive to give Skeedaddle and Lulu a look. He could tell right away they'd make terrific racers, but he wasn't happy about the price. Ken handed me a hundred-dollar bill, and when I repeated that the price was only fifty dollars for both, he put his hands over his ears.

"Make sure you come to the races," Ken urged, as he was loading our last two goats onto his truck. Before I could point that out, he added, "You don't have to bring your own. Lots of folks bring extras. You can borrow 'em. Rent 'em. Buy 'em. Yup, I bet you bring a few home. Once you see what goes on, it gets into your blood."

"Hell, no," I muttered to Mika. The second Ken's truck was off the property we'd be goat-free at last, and I had no intention of ever making that mistake again. We'd do just fine, thank you, with our handful of peaceful, home-loving, fence-fearing sheep, and no amount of goat race glory would ever be worth the headache of chasing another Bamboozle off the road at five in the morning.

Still, we were curious to see what a goat festival was like, so when race day rolled around a few months later, Mika went with the girls—

And came home a changed woman.

I'd thought we were on the same page with our NeverHorns policy, but that day at Falmouth opened her eyes to a whole new possibility. In one of the races, a little girl was sprinting toward the finish line when her goat suddenly dropped to the ground like it had had a heart attack. The goat lay there, stiff and motionless, while the little girl waited patiently by its side. After a minute or two, the goat stirred and came back to life. Together, they jogged the last few yards to the finish line.

"That's some hardcore fainter," one of the spectators told Mika. There's an entire breed, known as Tennessee Fainting Goats (or "myotonics"), that passes along a genetic mutation that causes their muscles to lock up when they're startled. Sheep farmers like them for herd protection; if you put a few fainters out to graze with your valuable ewes, the fainters will drop to the ground whenever a wolf or coyote attacks, giving the sheep time to flee. But for backyard farmers like us, Mika realized, fainters offer two other advantages: they're easy to catch, and they can't coil themselves to jump fences. They'd be perfect for milking, Mika urged, *and* for racing—because, yeah, Ken was right: go to Falmouth once and you'll be hooked for life.

A few months later, the girls found a long red string tied to the bottom of our Christmas tree. They followed it across the living room, out the back door, and through the backyard to the garden shed, where two young fainting goats were waiting. They were snow white and only a few months old, and for reasons I no longer recall but completely endorse, we named them Chili Dog and Awesome Blossom.

I'd found them in October at a small farm in nearby Rising Sun, Maryland, and the farmer agreed to hold them for me until I could sneak them home and into the shed on Christmas Eve. Chili was such a timid little tyke, I had to put him on the front seat of the truck with me so Awesome wouldn't bully him on the drive home. Since then, he's more than grown into his own, sprouting a magnificent pair of curvy horns that he's not shy about thumping into any creature in the barnyard that gives him a hard time.

Mika was right and wrong about the fainters: they couldn't jump, but every once in a while they'd get a taste for open spaces and Houdini their way out from under the tiniest gap beneath the fence. I could forgive them, though, because they really were perfect for the kids to race. We became Falmouth regulars, and no matter how hectic things got on race day, Ken was always easy to spot. Even at age seventy-five, he liked to arrive early so he could park right next to the starting line and be on hand to help when the youngest children were nervously leading their goats to the line for the very first time. Ken would welcome the crowd with his signature greeting—"Fine people of Falmouth! Are you ready?"—and then hand the microphone to his grandson, Nico, who'd taken over responsibility for play-by-play color commentary.

In 2016, something odd happened. For the first time in nearly forty years, race day arrived but Ken didn't. When we'd met Ken a few years earlier, he'd never let on that he was battling cancer. He fought it hard for twelve years, long enough for him to handpick the goats that a fourth generation of Brandts would run with. He hoped to see them race at the fortieth anniversary, but he didn't make it. This time, the fine people of Falmouth weren't ready.

Ken would never know the favor he'd done for Sherman. If it hadn't been for Ken and his Go Goats! attitude, Sherman wouldn't have had a goofy brown friend waiting to nuzzle and reassure him after he was rescued. You can draw a line directly from Ken's arrival at our house, to Chili, to Lawrence. But this morning, we were circling that line back to Chili. Overnight, my dozing brain had hit on something my waking brain had been too impatient to grasp: Anything I demanded from Sherman would never happen. Anything I offered had a chance. The more I tried to boss him around, the more he would resist. There was no way I was going to out-donkey a donkey, especially not a damaged one like Sherman. Years of captivity had seasoned him into a hardened resistance fighter who could outsmart, outlast, and outmaneuver anything I threw at him. For Donkey Tao to work, I couldn't push; Sherman had to pull.

But that didn't mean I couldn't dangle a carrot. Or in this case, some Chili.

I found Chili grazing behind the brown barn. He tried to bolt when he saw me approaching with a leash, but his back legs seized up, fainting-goat-style, and I was able to harness him before he regained full power. I led him back toward Mika, who was doing her best to control Sherman and Lawrence at the same time. I squeezed Chili past them and out through the gate, then went back for Sherman. Sherman followed Chili without a problem, probably because he hadn't yet realized that Lawrence was playing no part in this operation.

Chili had Falmouth experience,* so he'd become a good walk partner. Five minutes of walking with Lawrence, on the other hand, could leave you facedown on the pavement or wrapped in your leash tight as a mummy. Lovable as he is, Lawrence is a hyperkinetic freak, and the world is far too full of marvels and joys for him to ever do anything except lunge and swerve toward whatever catches his eye. Focus and impulse control aren't among Lawrence's strengths, or even his weaknesses. They're not even trace elements in his atomic matter. That's why we'd never brought him to the goat races; sure, he'd be fast, but whether that speed would take him toward the finish line or some poor kid's corn dog, we had no way of knowing—or controlling.

Mika held Lawrence back as Sherman went through the gate, then quietly turned Lawrence loose and shooed him away. Luckily, Lawrence spotted some lambs butting heads in the neighboring meadow and charged off to join them before Sherman noticed he was gone. Slowly, almost casually, Mika led Chili toward the road.

* Remember Johnny, the little white goat that Nancy Sweigart borrowed for her granddaughter to run with? Chili is Johnny's dad. Johnny was a spare goat I'd brought to the races that year.

I stayed back, holding my breath, feeding out rope to see what Sherman would do.

Suddenly, Sherman sensed he was alone. His head jerked up and began swiveling. In the distance, he could see Lawrence, off thumping noggins. Close at hand was Chili, just starting to walk away. Sherman paused, then began following Chili. When he reached the asphalt, he kept right on going, just like yesterday.

"See if you can speed Chili up when you reach that puddle," I called out to Mika.

"What puddle?" she said, looking around. She hadn't been with us yesterday, so she couldn't quite comprehend just how barely a puddle the Barely-a-Puddle of Doom really was.

"*Stain*, I mean," I was trying to keep my voice calm, but we were nearly on top of the thing and I was desperate to see if we could keep Sherman moving and avoid another open-road tug-of-war. "See that stain? No, that—"

By then it was too late. Chili meandered along, moving so pokily that Sherman had plenty of time to see through my trick and realize that the same terrifying shadowland he'd endured the day before was going to swallow him up in just 3, 2, 1 . . .

And we were through.

"Amazing," I said. "You should have seen the fight he put up the last time."

"He seems happy today," Mika pointed out. "Look at the way he keeps shaking his mane."

She was right; every once in a while, Sherman would waggle his head as if shooing flies, but there were no bugs around. It was more like he just enjoyed the feel of the breeze. I'd been so focused on troubleshooting, I hadn't noticed how different he looked, with his head high and his eyes searching around the road with interest instead of suspicion. As we turned off the pavement and onto the gravel road, Sherm quickened his step and pulled next to Chili instead of trailing behind. Together, the four of us walked side by side down the gravel road and into the woods.

"Tanya's right," I said. "He's like a prisoner who's been in soli-

Chili Dog leads Sherman on his first triumphant creek crossing.

tary. Now he's out in the world and everything looks weird and dangerous. But if he sees something once, he figures it out and remembers."

Alongside the gravel road is a creek, and after about a quarter mile, there's a dip in the roadside that makes the creek easy to access. When we got there, Mika led Chili down to the water. I knew we were throwing a lot at Sherman, but I had a reason: he was going to confront scary new things everywhere he went, so maybe it was better to forget long distances for now and focus on one scary thing at a time. Colorado's white water was going to be quantum forces more intense than this, anyway, so the sooner we started, the better.

We'd never had a reason to take any of the goats into the creek before, but Chili was a champ. Mika plunged in and Chili followed, wading through the shallows and scrambling over stones like a wild ibex. Chili was having a blast, but Sherman's entire body had turned into a living neon sign flashing a single message: *You've GOT to be kidding.* He'd come down the bank willingly enough, then stopped at the water's edge. I stepped into the current and pulled the rope taut, applying pressure, but I could tell we were done for the day. There was nothing in the Eastern art of gentle persuasion that would ever convince Sherman to give it a try.

Let him pull you, I reminded myself. The rope was long enough for me to hold on and still reach Chili and Mika, who were now on top of a midstream boulder, so I waded toward them, leaving Sherman on the bank. Behind me, I heard a clatter.

"He's going for it!" Mika called. "C'mon, Shermie!"

I was dying to turn and see, but checked myself. I was afraid if I looked, I'd ruin whatever was going on back there. The clattering and splashing got louder, then a snout thumped into my spine. I kept walking, eyes front, doing my best to pretend it was no big deal. Only when Sherman climbed up on the boulder beside us did we descend on him, scratching his head and ruffing those long ears, letting him know what an amazing leap he'd just taken.

10

Bag Man

Tanya had a soft spot for Sherman, but not that soft. She knew better than I did what I was asking from her with this scheme of mine—and from Sherman. Tanya had a farm and a business to run, and no time to coddle—let's be real—a bumbling amateur who was going to plunge everyone around him into a world of injury, frustration, and failure. When she came by to give me the news face-to-face, I knew right away what she'd decided. The trailer was a dead giveaway.

"Behold!" she said, sliding out of her truck and opening her arms, showman-style, toward the stock trailer she was hauling. The trailer shook with thumps and thuds, and then Tanya threw open the latch and out strolled Flower, Tanya's big riding donkey. Compared to scruffy gray Sherman, Flower was a breed apart: she was tall and athletic, with a chestnut coat as glossy as a mink stole and white rings around her deep brown eyes that made them glow as if she'd just emerged from hair and makeup for her cover shoot.

"Wait till Mr. Sherman gets a load of *her*," Tanya said. She walked Flower over to the gate and turned her loose in the little meadow. The sheep edged away, not sure what to make of this horse-sized stranger, and Sherman followed their lead, using the lambs as a buffer between himself and this curious thing he'd never seen before: another donkey.

"That's what I thought," Tanya said. "He doesn't even know what donkeys look like. Let's give them some time to sniff each other out." We went inside to have a coffee while Tanya explained her brainstorm.

The day before, I'd told her on the phone about the Chili Dog Initiative. Tanya was impressed, but after mulling it over for a while, she realized I was missing the big picture. Creek or no creek, goat or no goat, there was one crucial lesson that I could never teach Sherman on my own: how to be a donkey.

Sherman had grown up alone, so he'd never been part of a herd that could teach him how donkeys behave. Instinct will take you only so far; after that you rely on fellow creatures as role models. Sherman had never learned Basic Donkey, and it showed: just two minutes ago, we'd watched him literally becoming sheepish. Lawrence and Chili and the rest of his new barnyard family had really stepped up as companions, but what Sherman needed now was a different kind of playmate, one that could teach him the donkey equivalent of horseplay. If I wanted Sherman to trot across the countryside like a natural-born donkey, he needed to learn from one.

"So you're on board?" I asked.

Tanya stuck out her hand. "Let's go to Colorado," she said, and we shook on it.

Ultimately, the Man-on-Mars challenge was just irresistible. Tanya could play the whole chess game out in her imagination, seeing every move I needed to make in her mind's eye, but unless she was there to show me, there was a good chance I'd get things wrong and she'd never know for sure if it really was possible to transform sick Sherman into a long-distance racer. Plus, Tanya had a power card she could play anytime she felt Sherman was in over his head: she was the only person I knew with a horse trailer. I'm sure that somewhere on this boundless earth there's another volunteer who could drive me and a donkey six thousand miles round-trip from Middle-of-Nowhere, Pennsylvania, to Miles-from-Anywhere, Colorado, but who that person was and how I'd find her, I had no idea. If Tanya and I didn't see eye-to-eye on Sherman's training, she could put the freeze on this operation at any point by pocketing her truck keys and walking off the job, leaving Sherman and me stranded at home halfway across America from the starting line.

"Let's check how everyone is getting along out there," Tanya said.

We headed to the meadow and found that in the ten minutes we'd been gone, Flower and Sherman had decided to audition for their own segment on *Modern Love*. Granted, it was donkey-style, which meant a lot of mock kicking and snapping teeth at each other's necks, but Tanya pointed out that no matter how much it looked like fighting, both donkeys kept circling back to each other. Switch this scene to a high school, and we'd be watching two flirty teens thumb-wrestling in the lunchroom.

Tanya grabbed her saddle from the trailer while I got Sherman haltered and roped. I turned to help Tanya with saddle, blanket, and bridle, but even though Sherman is tall, Tanya is not, and all that stuff is heavy. Yet she heaved it around so fast that she finished as quickly as I did. I didn't know how we'd get not one but two donkeys through the gate without Lawrence hurtling past us, but Flower made that job a lot easier: Lawrence wasn't too keen to tangle with those big hooves, so he kept his distance. All of Sherman's nervousness, on the other hand, had vaporized; when Tanya led an obedient Flower down to the road, Sherman latched on behind.

"This is for you," Tanya said, handing me a riding crop with a plastic shopping bag knotted on one end. "Your own donkey guidance system."

Today, Tanya was planning to test Sherman out with some real donkey skills. And that, she explained, would require a lot more from me than just dragging a goat down the street in front of him. Historically, donkeys lead and owners follow. Donkeys like to be in front, because their No. 1 survival instinct is to scan the world ahead and make their own decisions, step by step, about where to place each hoof. That's one reason mountain men and other wilderness wanderers out there in the crazy places love donkeys so much; *you* might get drowsy, your horse might blindly follow orders, but donkeys are super-vigilant and will slam on the brakes, hard, whenever a patch of trail seems sketchy or a harmless-looking stick turns out to be a rattler.

Everything I'd been doing with Sherman, in other words, I'd been doing literally ass backwards. Sherman should be out front, not facing my back. We were able to get away with it while we were slowly trudging along, but if I ever expected Sherman to ramp up to a run, I had to give him the freedom to pick his own path. Otherwise, we'd be the worst combination of running partners: overcautious meets overcontrolling. We'd be two enemies bound by a single rope, waging a battle as he challenged where I was going and I wondered what he was doing back there and why he wasn't following.

The alternative? "Ground driving," Tanya explained. "You get behind him and steer from the rear." That's how the Amish train a young horse, she said; before you hitch it to a buggy, you stay on the ground and walk behind it, teaching it to respond to the reins and to voice commands. For Sherman, instead of reins I'd be using Tanya's bag-on-a-stick. We'd be going wireless: if I shook the plastic bag next to his right or left eye, Sherm should turn the opposite way.

"*Should*," Tanya emphasized. "Or he might bite it out of your hands and snap it in half. We'll see."

Tanya swung up into the saddle and clucked Flower into a walk. Without thinking, I began to pull Sherman along behind them. Tanya reined up.

"Really?" she said. "Already?"

"Right," I apologized. "Sherman goes first. Got it."

"Maybe I need a bag-on-a-stick for you too."

Tanya wheeled Flower around behind Sherman and began herding him forward. "You keep still," she told me.

This time I stayed put. Sherman and Flower walked on while I waited, playing out the rope until it was completely uncoiled and about to jerk out of my hands. Then I fell into step behind everybody, trailing at a distance. Tanya kept driving Sherman along until she felt that he—and I—had the idea.

"Now we're going to switch," she told me. "I'm going to push out ahead and you're going to move up here—" She pointed toward Sherman's haunch, right where she'd told me I was in danger of being kicked. "This is your sweet spot."

"My sweet spot? You said never stand there."

She pointed straight at Sherman's butt. "No, *that's* the kick zone. This"—her finger moved six inches to the left—"is your sweet spot."

I looked to see if she was goofing, but she was all business. Donkeys have terrific peripheral perception, Tanya said; they can rotate those big eyes backward and tune those antenna on the tops of their heads in any direction, allowing them to run forward while still keeping tabs on every movement to the rear. It's a great early-warning system against creeping predators, Tanya said. And for our purposes, it's also the perfect tool for ground driving; by positioning myself just outside Sherman's leg range, I could give him free rein to run while still signaling directions from behind.

Tanya trotted past Sherman and took point. Sherman jogged to catch up, and that's when I saw my chance. While Sherman was focused on Flower, I hurried toward his butt and stepped into—maybe?—my sweet spot, somewhere that I hoped wasn't too left, too right, too close, too far. Sherman's left eye fixed on me, but he kept his head forward and continued trotting. Tanya and Flower led us past the Barely-a-Puddle, and then we turned onto the—

Oof! I slammed into Sherman, who'd suddenly stopped dead in his tracks. He jumped when I sprawled into him, coming down hard on my foot with his hoof. Pain sucked the wind out of me, leaving me too breathless to speak and driving out every thought except *Get this goddamn beast off me!* I managed to shove Sherman aside and free my foot, then sucked in air, trying to bite back my seething.

"Flower!" Tanya scolded. "Big baby." Flower was backing up warily, her face inches from the ground like a bloodhound on the scent. Tanya urged her forward, but Flower wasn't having it. She walked left, then right, snaking back and forth like a drunk getting street-tested by a cop. "Sorry about the pileup," Tanya said, still trying to wrestle Flower back on course.

"She's worse than Sherman about water," Tanya explained. That's for sure; even Sherman had walked past this spot without a problem yesterday, never noticing that a tiny stream trickled under the road through a culvert buried two feet underground. Tanya got Flower going again, but those few moments of magic we'd just enjoyed seemed to be over. Flower had recovered from her little panic attack, but Sherman hadn't: he'd seen enough for today and turned for home. I tried hauling back on his rope, but he was determined.

"Tanya, little help," I called.

Tanya wheeled Flower around and came back. She got Sherman turned in the right direction, then trotted Flower forward to lead. I crowded in behind Sherman, tucking myself into that sweet spot and trying to block him from U-turning again. Sherman must have sensed he was surrounded: this time he set off smoothly. We jogged along beautifully, both of us comfortably in stride. Sherman drifted a little to his left, so I did too, happy to get out of his way as long as he was moving. He edged over more . . . and a little more, until I realized I was nearly squeezed against the side of the road. I jumped up on the grass embankment to avoid getting pinned, and that's when Sherman pivoted and bolted for home, tearing the rope from my hands. The little sneak had conned me.

"I got him," Tanya called. She'd been watching over her shoulder, and once again she rode to the rescue. She cut Sherman off, then slid out of the saddle. I grabbed the rope and reached in my pockets for some treats, figuring she was going to reboot him with a little soothing, but instead she took Sherman firmly by the halter and turned him around in the right direction.

"Nope, we're not bribing him," she said. "We don't reward him for quitting his job."

Then she turned to me. "That was your fault. You quit first."

"No, he—" I began to argue, then saw that she was just getting started and shut up.

You're not his dictator, she began.* You're not his slave driver. You're his *leader*. Sherman has watched out for himself his entire life. He's not used to relying on someone else, and he's never, *ever* going to rely on you unless you deserve it. Seriously, why should he follow your orders if you're not paying attention? That's how a herd works: it doesn't matter if you're in the lead or bringing up the rear, you have to prove you're on the ball. Donkeys operate on one frequency—trust. They do nothing on faith, but everything on certainty. They can be dying of thirst, but if they're not sure about the water, they won't touch it. If the hay smells iffy, they'll go hungry. And if you're not covering their flanks, they'll do it themselves.

So that trick Sherman played back there? Tanya continued. He tested you and you bombed. The herd leader has to anticipate trouble and avoid it. Instead, you moseyed straight toward the rake and stepped on it. For a donkey, a mistake like that can be death. In Sherman's world, a wolf lurks behind every bush, a mountain lion crouches in every tree. You're teaching him that his job is to run past those bushes and under those trees. Well, what's your job? Aren't you supposed to be watching out, protecting him?

"That thing in your hand," Tanya said, meaning my bag-on-a-stick. "Use it. Take charge. Show Sherman you know what you're doing." She swung back into the saddle, then added a final tip: "When Sherman knows he can trust you, it will change everything. You'll see a difference in him you won't believe."

* Or words to that effect. I didn't catch the exact phrasing of Tanya's speech, but I certainly remembered the heat. This is a pretty good approximation.

. . .

Tanya meant what she said about taking charge. Instead of herding Sherman forward, this time she walked Flower around us and kept going, leaving me to figure things out for myself.

"Walk on," I told Sherman, echoing the command Tanya used for Flower. "Let's go. Walk on."

Sherman was just as no-nonsense as Tanya. Paying no attention to me, he turned toward home. I jumped in front and headed him off.

"Nope," I said, spreading my arms to block him. "We're going to walk on."

Sherman paused. I stepped toward him, arms still wide, crowding him so he had only one choice: turn around or knock me over. He edged right, then left, and so did I, stepping closer each time. Sherman backed up . . . then slowly pivoted. Ahead, Flower had eased into a trot. Sherman's head jerked high, as if noticing for the first time that she was gone, leaving him in no-man's-land—home far off in one direction and his new friend about to disappear in the other.

"Let's get 'em, Sherm. Now *walk on*!" Sherman lurched forward, trotting so briskly that I almost lost the rope again. Flower and Tanya were a few hundred yards away and going strong. I was afraid Sherman would give up when he realized we couldn't make up the gap, but as long as Flower was in sight, he stayed in pursuit. Only when Flower vanished down a dip in the road did Sherman rebel. He immediately spun for home, but I was ready for him. Before he finished his U-turn to the left, I had my bag-on-a-stick in front of his eye. Just as Tanya predicted, he turned back right. He kept on turning, but the bag was waiting for him that way too.

Boxed in, Sherman stopped to consider his next move. I took him by the halter, pointed his head down the gravel road, and let him weigh his options. Tanya and Flower hadn't come back, so it was just the two of us, locked in a stalemate.

"Walk on," I ordered, not knowing what else to say but fully

Sophie and Sherman help Tanya work on Flower's water phobia.

aware that I had just as much chance of Sherman obeying if I'd ordered him to sing Happy Birthday.

"Walk—" I tried again, and before I'd finished, Sherman was on the move. He took off at a nervous trot, head high as he searched for Flower. I was afraid he'd give up and start fighting toward home if we didn't find her soon, but we rounded a bend and spotted her ahead. Sherman suddenly veered sideways, as if all he'd wanted was confirmation that Flower was alive before retiring back to the barn, but I had my plastic bag on hair trigger and flicked it up in time to straighten him out.

Luckily for us, Flower had met her perfect storm: not one but *two* creeks, one of them gurgling beneath the road and the other cascading into a frothy little waterfall to our right. This was the hill she'd die on, Flower had decided, so Tanya had her hands full as Flower zigzagged and backtracked and stalled. When Tanya saw us coming, she decided to take a break and wait for us to catch up.

"That's how, cowboy!" Tanya hollered as we approached. "That's the way to ground-drive."

We all paused beside the waterfall. Tanya wasn't going to let Flower finish the day without conquering that trouble spot, so she kept working till she'd coaxed her past it. Sherman tromped along contentedly, not bothered by the water as long as it wasn't directly underfoot and Flower was nearby. He was so relaxed, it took a while before I looked around and processed what he'd just done: the sick, lame donkey marked for death last week had run half a mile.

"Wow," I told Tanya. "If we make it home from here, that's a full mile. Unbelievable."

Tanya looked Shermie over, taking in his easy saunter. "I think he's ready to show us more," she said. "Let's push on and see what he's got."

11

Wild Thing

One month later, the four of us were once again stalled in the same spot. "Know what I love about Flower?" I told Tanya. "She still acts like she's never seen that waterfall before."

Day by day, Sherman kept getting stronger because Flower, conveniently, wasn't getting any braver. Tanya and I had settled into a schedule of three runs a week, and every time, the two donkeys would follow the same pattern: Sherman would futz around at the beginning, testing me with his twists and feints, until he'd suddenly realize Flower was gone and get down to business. He and I would push hard to catch up, usually a little too hard, hammering the gravel road at a clip that was uncomfortably quick for both of us. At about the point when Sherman was losing hope and I was sucking air, there we'd find Flower: pawing and snuffling the ground like a kid peeking over the edge of the high dive, suspiciously sizing up the same patch of road next to the waterfall that, by the end of four weeks, she'd crossed more than fifty times.

Not even the UPS guy was surprised by us anymore. Over the past month, he'd come across us enough times to know that whenever he was near AK's Saw Shop, he'd better keep his foot hovering near the brake. "Only down here!" Tanya would holler as he slowed to squeeze between us and the creek. "Only in the Southern End."

Personally, I loved Flower's weird little Rain Man–nerisms, because they gave Sherman and me a chance to regroup and recover before pushing on with the workout. Without those time-outs, Sherman might have quit before the runs really got started. And it wasn't just water that freaked out Flower; she also sensed the hand of death looming in:

- tire skid marks
- cracks in asphalt
- bridges of any type
- a scrap of pink survey tape hanging from a tree branch
- a curve in the road that was too sunny
- a curve that was too shadowy
- shadows in general
- the color yellow, especially in road signs and underground-wire warnings
- cows (But not dogs. Let a farm cur come raging out of the barn, and Flower will yawn. Let a friendly heifer sidle up to the fence, and Flower bolts.)

To this day, some Flower triggers remain an utter mystery. Even Sherman would be baffled; we'd all be jogging serenely through the woods, nothing around us but trees, when Flower would freeze so fast that Sherman would run right by before realizing something was supposed to be scaring him. The three of us would stand there, baffled. "Ah, *there* it is," Tanya might say, finally spying a hunter's camouflaged deer stand fifteen feet up in a tree. Otherwise, we'd shrug and move on, chalking it up as a menace detectable only by FlowerVision.

Sherman's fixation went the opposite way: instead of fleeing life's little surprises, he followed them. That sank in one afternoon when Tanya wasn't available and the girls and I decided to take Sherman out on his own. We'd gotten so used to Sherman's excitement at chasing Flower the half mile down to the waterfall, we'd forgotten that flying solo might be a very different experience. Sherman started off strong, happy to be sandwiched between me and my daughters as we jogged down the driveway and into the street. We turned onto the gravel road, and only as we were approaching the waterfall did Sherman wake up to the fact that this time Flower wasn't going to appear. He moped and balked, intent on turning us around for home.

"How about we give that a try?" I suggested, pointing off-road toward a dirt trail near the waterfall. The trail winds up a steep hill and through the woods, eventually emerging from the trees at a small farm with two big, fiery horses. We'd never tried it with Sherman because the climb is tough and the horses are tougher; true, they're fenced in, but they love to greet strangers by thundering across the meadow like demon steeds of the Apocalypse and skidding to a stop, nostrils flaring, just inches from the wire. They're fun to watch, though, and I didn't think Sherm was really at risk because if we couldn't persuade him to walk any farther down that flat road, there wasn't much chance he'd grind all the way up the hill to the farm.

But something came over Sherm when his hooves touched that dirt. It was an electric reaction, as if power was surging into his body from out of the ground. Sophie and Maya were scrambling ahead of him up the hill, but Sherman's four-wheel drive blew right past them. I dropped the rope and the girls got out of the way, surrendering the trail to this suddenly inspired donkey who churned around the switchbacks. About halfway to the top, he stopped and looked back, apparently as surprised to find himself out front as we were. For the first time, he could see the world from the head of the pack. He must have liked what he saw because he shook his mane and climbed on, speeding up the rocky trail and disappearing over the crest.

Uh-oh.

"Is he coming back?" Maya asked.

"Um . . . yeah. Sure!" I lied. This was an entirely new problem. Until that moment, I'd never had to worry about Sherman running too much. He'd been back on his feet for only a month, and during that time he'd never been voluntarily separated from his nearest friend by a distance of more than six inches. Bolting like a runaway racehorse was never a scenario I saw coming. I felt a surge of joy—*Run free, buddy!*—followed by the heart-clenching fear that for all I knew, Sherman wouldn't stop galloping until he plunged in front of a car.

The girls and I hiked up the trail as fast as we could. And there, transfixed, was Sherman.

He'd come to a stop at the edge of the hilltop farm. Through the fence, he was touching noses with the two big warhorses. All three seemed fascinated; Sherman by these magnificent cousins of his, the magnificent cousins by this scruffy gray creature who'd just erupted from the woods and was now sniffing them over, awed but unafraid. We gave Sherman time to bond with his new heroes, then took him by the rope to head home. As soon as we were back on the trail, his dirt-induced superpower kicked in again. I couldn't fight him and keep my feet, so I chucked the rope and let him go, watching as he descended the hill like a slalom racer, nimbly picking his way through the rocks and around the trees.

I slipped and scrambled after him, knowing I'd have to run hard once I hit the gravel to have any chance of catching up with Sherm before he reached the main road. Turned out I didn't have to worry; when the girls and I made it down, Sherman was right there by the waterfall, waiting.

It was a great day, but a confusing one. Was it just the dirt that got Sherman so fired up? And why did he stop at the bottom when he could have motored all the way back to Lawrence and Chili? I couldn't figure out what Sherman was thinking until that evening, when a cat spoke up and gave me a clue.

I was out collecting the eggs from our chicken coop when I heard desperate meowing. I hunted around and found Polly, a calico kitten we'd taken in, being stalked by Sherman. Polly had scampered to the top of the split-rail fence to get away from him,

Sherman gets the game, while adopted stray Polly gets outta there.

but Sherman wasn't giving up; he kept nudging her with his nose as Polly tried to tightrope-walk her way to safety. Sherman looked so happy, I'd swear he was actually grinning. I pulled out my phone to take a few pictures, but then I had to jump in and spoil his fun when Polly popped out her claws to julienne his nose. I swooped between them, carrying Polly to the house while Sherman lumbered along behind us.

Inside, I checked my photos. The short sequence showed Polly transitioning from confused, to annoyed, to downright angry. But it was the expression on Sherman's face that riveted me. Something about it looked oddly familiar—

And then it clicked. *Holy shit,* I thought. He gets it. He gets the game.

In the photos, Sherman has the same delight in his eye that the girls and I had seen when he gazed back at us on the trail. He wasn't trying to flee; he was trying to *play*. In Sherman's mind, I realized, our runs with Flower were a cat-and-mouse game, and he was the cat. It was up to me to give him something to chase, but today, he reached the waterfall and discovered I'd messed up: no mouse.

So what did Sherman do? He found his own. I don't know if he detected those horses by their scent in the air or the tracks of hooves but he sallied after them the same way he usually pursued Flower. He'd glanced back on the trail to see why the girls and I weren't keeping up, but as far as he was concerned, that was our problem. What, was he supposed to wait around forever? There was quarry to be caught! When we got home afterward, Sherman was so keen to keep playing that poor bewildered Polly found herself cast against type, not to mention her will, as the new rodent designate.

It all made sense. I'd taken a game—the burro race—and turned it into a job; Sherman had taken the job and turned it back into a game.

Now it was up to us to make him good at it. It didn't take me long to realize that Sherman's game had one glitch: the cat and mouse liked each other too much. Whenever Sherman caught up with Flower, they'd be so happy to see each other that when it came time to run again, neither wanted to go first. Tanya and I would be spinning in circles before we could straighten the lovebirds out and get moving. I was glad to discover he was actually having fun, but catching Flower wasn't going to get Sherman to the finish line of a mountain ultramarathon, not with all that standing around we were doing. We had to figure out a way to keep things playful but add a little urgency and ramp up the running. Compared to our gentle, three-mile jaunts along the creek, the World Championship Pack Burro Race was going to be a far more ferocious beast.

Tanya and I needed some help, and I had an idea where to look. It was time to call in Vella Shpringa—the world's only Amish running club.

12

The Zipperless Guide to Better Living

When Amos King gave running a try, it wasn't as if his legs weren't already getting blown out every day. He was a twenty-six-year-old roofer, which meant climbing ladders under a hot sun with a stack of asphalt shingles on his shoulder, and because he's Amish, just getting to work was a workout. Most mornings, Ame was out the door before the sun was up, pushing himself down the road to his shop on a heavy steel scooter that required so much force, he'd soon be panting and switching from right foot to left and back again to relieve his burning quads. He'd end his days the same way: while his non-Amish crewmates were cranking the AC in their pickups and digging in the cooler for a cold one, Ame was scooting on home, powered by his own sweat-propulsion engine.

Ame is strikingly handsome, with the kind of confident, genuinely friendly appeal that grabs your eye as soon as he enters a room. But back then, one secret fear nagged at his self-esteem. Even though he was at the age when he should be settling down, finding a wife didn't worry him: he was afraid of what would happen next. "My initial reason for running was to not get fat," he says. "In Amish culture, as you get older and married, and with all the good cooking, you're doomed. You're *doomed*, brother. I didn't accept it. You create your own destiny."

Ame decided it couldn't hurt to try a little extra exercise, so when one of his buddies from work mentioned he was signing up for a 5K, Ame asked if he could come along. Race day was freezing cold, and Amos wasn't sure how heavily to dress or how hard to push. "Some guys who'd run before told me, 'Don't go out too fast,'" he recalls. "Guess what Amos did?" The leaders shot off at a sub-five-minute-mile pace—and right on their heels came the Amish roofer in long black pants and suspenders, attempting for the first time in his life to run three consecutive miles in a row.

"It felt easy," Amos says. "At first. Then my nose was running and my nostrils froze together. I couldn't breathe. I had to walk. It was hilarious."

Nevertheless, he finished in twenty-two minutes, a performance respectable enough to please most recreational runners, but not quite good enough for Ame. He was chatting about running with his insurance agent, who put him in touch with Jim Smucker, a veteran marathoner and third-generation owner of the Bird-in-Hand Family Restaurant & Smorgasbord, an Amish country institution in Lancaster County. Ame was hoping for some tips, but Jim did him one better: he invited Ame to join him and a friend for a speed workout.

A few evenings later, Ame met up with the two Mennonite marathoners at the Conestoga High School track. They were going to do 800-meter intervals: two laps at top speed, followed by a jogging recovery, repeated over and over until you wish for a merciful death. Jim warned Ame that it was a killer workout, so he shouldn't worry about keeping up. After nine intervals, though, Ame was still right on the heels of the veteran runners. In fact, he didn't even seem to be struggling.

"Don't be polite," Jim said. "Why don't you open 'er up on this last one and see what you have left?" Permission granted, Ame blasted off. He sprinted so hard, he opened up a lead of more than 200 meters and finished ahead by a full thirty seconds. "Their jaws just dropped," Ame told me. "That's when they thought, 'Oh, I think the Amish guy *does* have some speed.'"

True, Jim was impressed, but he knew he'd never see Ame again. Nothing against Ame, but Smuckers have been in Lancaster for more than a century, and in all Jim's years, he'd never even heard of an Amish runner before. And for good reason: it all boiled down to a spat that broke out between Jim's and Ame's ancestors three hundred years ago.

Originally, Amish and Mennonites were a single faith led by Menno Simons, a maverick Christian in the Netherlands who believed that infant baptism was bogus. How do you build a committed, worshipful community, Menno asked, when you're dragooning babies into the ranks who don't even know what's happening to them? Menno's challenge didn't sit well with the dominant Catholics and Protestants, who felt that perhaps they could bring these strays back into the fold through the warm embrace of torture, murder, and terror. The persecuted Mennonites fled into neighboring countries, like Switzerland, where they did their best to adapt and fit in.

Fitting in, however, began to rankle Jakob Ammann as much as font-dunking newborns bothered Menno Simons. Ammann was a Swiss tailor who never learned to read or write, but he'd absorbed enough of the Bible and Mennonite teaching to believe that if Jesus stood for anything, it was for *not* fitting in. Ammann and his followers—the Amish—broke away from the Mennonites, creating a Plain People community which, three centuries later, is fundamentally unchanged. Today, Mennonites in our area drive whatever they want and, while still dressing modestly in long pants and dresses, are fine with jeans for men and pretty patterns for women. But Old Order Amish remain largely frozen in time, wearing the same style straw hats and mortician-like outfits as their great-great-great-grandparents' and relying on the strength of their own bodies and their kinship with animals to raise the food they need to survive.

It's a tough life, and it starts early; whenever I grumble in winter waiting for my old Ford to warm up, I quiet down as soon as I look up and see the parade of Amish kids hiking through heavy snow in the cornfield behind our house on their long walk to school. One frosty morning, I picked up a neighbor at five thirty to go to the Amish hardware store, and his three preschool-age sons were already working in the barn with him.

"Are they always up this early?" I asked Daniel.

"If they want breakfast, they are," he said.

I didn't really get this insistence on rugged old-fashionedness when we first moved to the Southern End. It was charming, for sure; we loved sitting on the porch in the evening and hearing the soft drumbeat of horse-drawn buggies instead of car horns and the roar of city buses. But seriously—phones have to be in a shed in the cornfield, not in the house? Chain saws and in-line skates are fine, but bikes and Toro lawn mowers are banned? Teens can play dodgeball, volleyball, and ice hockey, but not baseball? No electricity, no tractors—*no zippers*? Was there some kind of Old Order Da Vinci Code at work, or was it just to make life harder than it had to be for no real reason, like deciding you'll write only with your left hand or back your car into a parking spot using just the mirrors? Every time I thought I was getting a handle on Old Order-liness, I'd run into another rule that seemed less about revealed truth and more about . . . well, I didn't really know. Was it thought control? Sexual denial? Straight-up crazy fundamentalism?

I was already perplexed, and that was even before I met Sam Stoltzfus and learned about his Nazi espionage library. I ran into Sam for the first time when I got lost on my way to a local cabinetmaker's house and flagged down a passing buggy to ask for directions. Sam pulled over, and when he told me he also was a woodworker, I hired him instead. I followed him to his barn so he could stable his horse, then brought him back to my house to take measurements for the desk and bookshelves I needed.

During the drive, Sam was curious to hear how an "English"*

* By now, everyone knows that the Amish refer to all non-Amish as "English," right?

guy like me had wandered into the River Hills from downtown Philly. Needless to say, I had a million questions for him, too. Like, how come one of my neighbors was shunned for using a tractor, yet the community turned out in force to support a pair of Amish teens who'd been arrested for what had to be the grand enchilada of all Old Order sins: they'd been arrested for dealing crack to other Amish teens, after narcotics officers began wondering why two young guys with soup-bowl haircuts were driving in and out of Southwest Philly to visit the Pagans motorcycle gang. How was that fair?

Sam not only understood; he added a few juicy tales of his own. Like the one about the cops who showed up at the scene of an accident and found an Amish boy passed out in a buggy with blood on the road but no horse: the boy was so blitzed, he never woke up when a hit-and-run truck driver slammed into the horse and carried it another quarter mile down the road before it fell off the hood. Sam and I were getting along so well, he invited me to come back to his shop and hang out while he explained Amish thinking to me. In return, he had a favor to ask: could I find a few books he wanted?

"I can try," I replied, figuring I might locate whatever prayer books he wanted on Alibris or the Gutenberg archive. "What are you looking for?"

"*Seven Habits of Highly Effective People*," Sam said. "And *How to Win Friends and Influence People*."

"Really?" I hadn't seen that coming. "Uh—in English, right?"

"Yes, English. And Nazi spy books. Can you find some of those?"

Self-help and storm troopers—wow. Sam explained that a few weeks earlier, he'd picked up a box of old paperbacks at the "mud sale," our fire company's outdoor auction held each year in a soggy field after the spring thaw. Sam wanted only the gardening and herbal remedy titles, but included in the box were also Frederick Forsyth's *The Odessa File* and some personal-improvement books. Out of curiosity, Sam dipped in and found himself enthralled. The self-help stuff I could see, since I'm sure he recognized a lot of himself in the continuous-growth/seek-first-to-understand ethos, but I wouldn't have expected him to come away with a heightened interest in Israeli revenge killings and West Berlin strippers.

"Won't you get in trouble?" I was happy to lend him my own Forsyth books, but I didn't want to be a coconspirator in my new friend's excommunication.

That was when Sam opened my eyes to what I'd been missing. Amish life isn't about what you can't have, he explained; it's about what you *can*. What are the three things every person wants? Health, happiness, and security. Right? Well, the Amish are happier, healthier, and safer than the rest of us. *By a long shot*. This isn't just opinion; it's hard math.

Let's start with health and security. The Amish don't go hungry, homeless, or broke, because they've quietly created their own little semi-socialist Scandinavian paradise right in the heart of red-state America. They adopt one another's children, care for their old folks at home, build and pay for their own schools, and take in the needy. The Amish dodged the whole subprime mortgage fiasco by avoiding banks and shady brokers, relying instead on "Amish Aid," a community pot that provides loans and homeowner's insurance. Likewise, the Amish essentially created their own universal health care a long time ago, paying out of pocket for medical costs and helping their neighbors cover hospital bills by raising funds through auctions, donations, and sales of chicken pot pie.

But of course they come into health care with one big advantage: better health. The Amish are *six times* more active than the average American, and half as likely to suffer from cancer or diabetes. They don't smoke, drink, fight, or mess with drugs. They don't get fat (the Amish obesity rate is a nearly nonexistent 4 percent versus almost 40 percent for the rest of us) or fool with guns (which kill or injure some 100,000 Americans *every year*). They're big on real foods—kombucha, raw milk, grass-fed meats, organic produce, gut-enhancing fermented veggies, home-baked bread—and seldom eat out. They age gracefully, with far superior late-life mobility and overall health. The Amish rarely harm themselves, or anyone else. Their suicide rate is 70 percent lower than ours, and there has been exactly *one* Amish murder in all of American history, and that by a delusional psychotic. Which poses a sub-question: why is Amish mental health so sound that in three hundred years, they've created only one threat to public safety?

Happiness can be tricky to measure, but if we look at the same metrics as used by the retail industry—customer loyalty and return business—Amish numbers are booming. You might assume that an eighteenth-century society surrounded by shopping outlets would be dying out by now, but the community has steadily doubled every twenty years, a growth rate five times higher than that of the U.S. population as a whole. The Amish have a better retention rate than Netflix: roughly 90 percent of young Amish adults choose to stick with the faith and join the church for life. And thanks to Menno Simons, they know exactly what they're doing; baptism occurs after Amish teenagers have spent a year or more on *rumspringa*, or "running loose." They're free to buy cars, dress English, fly to Disneyland, pound Jäger at Mardi Gras, and even be forgiven if they're boneheaded enough to buy drugs from Pagans. Once they've experienced the modern world, the overwhelming majority decide, *Meh, not so great after all*, and return to "living plain."

"We're not perfect people," Sam cautioned me. And tragically, every so often someone either in the community or affiliated with it will prove it. A few miles from our home, a couple who'd left the Amish church fourteen years earlier was convicted of giving nine of their daughters to a sex abuser who'd convinced them he was the "Prophet of God." Back in 2011, a splinter group of Amish extremists in Ohio began terrorizing other church members by ambushing the men and forcibly shaving off their beards. Rather than bringing in law enforcement, the Amish prefer to discipline* their own through public penance and shunning, and that has left some women dangerously unprotected. An Amish molester in Missouri was stopped only because the church reluctantly called

* If you've seen Amish Mafia, by the way, you've enjoyed a complete serving of fiction. No one has ever patrolled Lancaster County cornfields with shotguns or strong-armed errant Amish bishops.

the police after his third offense, and an Amish bishop not far from Lancaster was arrested after failing to report two cases of child molestation because he said it "wasn't really that bad."

When talking and time-outs fail, that's where Amish culture goes wrong. What's amazing is how often it goes right. One afternoon, Sam and I went to visit his cousins' buggy repair shop near Bird-in-Hand. "My uncle made a mistake with this business," Sam told me. "He earned too much money." Sam's uncle was a whiz at salvaging unfixable buggies, even ones that had been crunched by cars in collisions. Since new buggies can cost up to $10,000, he began getting work towed in from as far away as Indiana and Kentucky. Word of his skill even spread to Disney and the Smithsonian, which hired him to restore vintage Wild West carriages. Then, at the peak of his success, Sam's uncle hit the brakes. He gave away more than $1 million in savings, divided the business among his nephews, and moved his family to a small produce farm. Why?

"Raising his children rich wasn't fair to them," Sam explained.

And right there, in the moment that Sam's uncle closed up shop, you can find the secret to Amish success. Sam's uncle knew that happiness, health, and security come from devoting yourself to two things—your family and your friends—and anything that doesn't bring you closer to both is pulling you in the wrong direction. Distance and envy are two poisons that can destroy any community, and that's why the Amish have a problem with cars, fashion, and even electricity: they let you travel too far, show off too much, and stare at screens instead of faces.

Sam's uncle loved his craft, but he loved his community even more, and when he felt himself being drawn away by constant praise, easy work, and fat paychecks, he had to make a change. His decision was a declaration of faith in the five words that define Amish life:

Slow down. Savor your world.

Most of us whipsaw back and forth all day, racing to save time so we can sit around and waste it. The Amish are skeptical of speed, so before accepting any new technology, they question whether it makes life better, or just go by a little faster. They don't automatically reject new things; instead, each Amish district debates for itself whether this new thing will help them learn patience, self-control, and empathy. If not, maybe the smart play is to avoid it.

But even while Sam's lips were still moving, I was arguing with him in my mind. I was on board with his logic about TVs and cell phones and maybe even air travel, but c'mon: enough already with the buggies. I kept my mouth shut because I felt that challenging him on this point was cutting a little too close to his core beliefs, but think about it: if the goal was to spend more time on your land and with your family, savoring your friends and God's green earth, then what was the sense of creaking along in a black box for two hours because you needed to pick up a pound of flour? *Especially* when there was nothing stopping you from hiring a car whenever you felt like it. Sticking with the buggies was silly and stubborn, a pointless knee bend to a lost past—or so I thought, until a donkey that was afraid of puddles arrived in our backyard and suddenly the whole Rubik's Cube machine of Amish interlocking logic clicked for me.

It's no coincidence, I realized, that the only Americans who don't need cops, fists, or therapists to settle their differences are also the only ones who haven't abandoned their business partnerships with animals. Patience and kindness don't show up on demand; they're disciplines that require constant practice, and there is no better boot camp for learning those skills than hitching your survival to your ability to discern—and respect—the needs of another creature. My Old Order neighbors understood that horses are less about transportation and more about education; for every hour they devoted to training their animals, their animals were quietly returning the favor. If you wanted to yank out the one piece in the Jenga tower that could make Amish culture and character come tumbling down, it's easy: take away their horses, and watch centuries of fellowship and nonviolence begin to fray.

There's a lot I will never adopt from my Amish neighbors (long black pants in summer and a cap on education at eighth-grade spring to mind), but Sam opened my eyes to the difference between rules that hold you back and rules that help you grow. That was why he could read thrillers if he felt like it and had no qualms telling me that he'd seen (and kind of enjoyed) the movie *Witness*. Amos, our closest neighbor over the hill, dropped in one evening while friends were over for dinner. "This is wine?" he asked, never having seen it before. "Can I try?" Before I could reach for the bottle, he'd filled a water glass to the brim. He drank it off like it was lemonade, then set off to walk tipsily home in the dark. "Yeah, I don't think I'll be having that again," Amos told me the next day. "Not enough evenings in life to spoil another one." The Amish aren't closed to the world, he's saying; they're just a little more goal-oriented about how much of it to let in.

13

Full Mooning

The Naked Mennonite, though; that one had *Deal Breaker* written all over his bare chest.

On hot evenings, one of the Mennonite runners who invited Amos to the Tuesday night track sessions would shuck his sweaty shirt, just like most other guys, but most other guys aren't surrounded by 63,000 neighbors whose bodies are banned from exposure to sunlight under penalty of eternal damnation. Jim Smucker enjoyed watching Amos develop his remarkable talent, but he knew that getting spotted with the Naked Mennonite was bound to land him in serious hot water.

Jim had seen it happen in his own family. Jim's uncle was Old Order Amish, and as a teenager, he'd become a hell of a baseball player. Amish kids love the game, and just about every Amish schoolhouse has a chain-link backstop in the play area where the kids, boys and girls alike, split into teams during recess. Amish youngsters became such strong hitters and fielders that by the 1980s, word had spread to semipro leagues in Pennsylvania and Ohio that instead of recruiting in, say, the Dominican Republic, all they had to do was drive around Lancaster County and they'd find a literal farm system thriving among the cornfields. A few enterprising teams managed to sign these Amish fireballers and issue them uniforms, and that's when the hammer came down. As soon as photos of the Amish players began turning up on the sports pages, the Old Order elders decided things had gotten out of hand. Only kids could play baseball, they decided. Teens would be limited to pond hockey, volleyball, and "eckballe," an Amish dodging game using human targets and a rock-hard leather ball. There wasn't much risk of any Amish skaters, spikers, or eckballe players going pro, which meant they'd wear their own clothes, stay out of the papers, and compete for fun, not fame.

But Jim Smucker's uncle went rogue. He was a gifted pitcher who'd been signed to a team along with his best friend, a catcher. For years, the two young Amish men had to sneak away and play under fake names so no one in the church would find out about their secret second lives. They never got caught, but it forced Jim's uncle to spend a good part of his life ashamed of his talent, hiding the thing he loved from the people who loved him most.

Jim Smucker knew that if Ame pushed things too far with the running, he could bring the same judgment down on his head. Ame was still a single guy living with his parents, so he had a little leeway. But at twenty-six, he'd reached the age when it was time to settle down and decide whether he was in the church or out, and from the looks of it, running was so far out that it could qualify for its own circle of hell. How could Ame justify racing in a 10K, where he would not only be vying for personal glory but hanging around a flesh-baring mob of women in sports bras and men in sausage-hugger shorts? Running was solitary, speedy, and show-offy, a devil's playground ruled by the false idol Strava. And hanging out with a Naked Mennonite, of course, wasn't making his case look any better.

But Jim had learned one thing about Ame: when you think he's beat, that's when he's at his best. Maybe Ame couldn't convert the Amish to running, but what if he converted running to the Amish? What if he turned baseball into volleyball?

Like most dating-age Amish, Ame belonged to a youth group. On weekends, the gang would get together to hike and picnic, maybe go to the demolition derby up at "the Buck," and definitely, whenever the weather was nice, set up a net in a big pasture and divide into coed teams for hours of volleyball. Amish elders have no quarrel with volleyball, because so far, it hasn't gotten out of control: the youngsters wear their own clothes, stay out of the news, and compete for fun, not fame. So if you're Amish, single, and ready to mingle, you can't beat volleyball. It's the perfect way for an Amish gal to flirt with that cutie she's got her eye on by "helping" him with his serve, because everything is right out in the open. All Ame had to do was keep the spirit of volleyball and get rid of the nets, and he might be on to something.

When Ame's youth group went to Lancaster County Park one Sunday afternoon, he pitched them an idea: instead of hiking the trails, why didn't they just cut loose and run? "I didn't know what to expect," a member of the group named Liz would later tell me. "I don't know how the guys talked us into it." But yeah, she does: she kind of fancied Ame. Off they went, tearing through the woods, the guys in their long black pants and suspenders, the gals in their long dark dresses and starched white headcoverings, taking on the rocky trails in whatever sneakers, boots, or shoes they happened to be wearing.

"I was very surprised I could run four miles," said Liz. "I almost fainted on Rock Ford hill." She was even more surprised the next morning, when she woke up with blisters and aching calves but wanted to do it again. One evening after supper, she told her parents she was heading out for a run.

A what?

"My family thought it was ridiculous," Liz said. "I was brave and went anyway."

This all made so little sense that Liz's mother got a scooter and followed. Liz ran five miles that evening, with her mother scooting along behind her. "That was motivation for me, because I could run faster than she scooted," Liz said with a laugh. That became their evening ritual, a quiet hour for Liz's mom to glide along, away from the house chores and her eleven children, watching her strong, brave daughter fly over the roads faster than two wheels could keep up.

The gang called itself Vella Shpringa—Pennsylvania Dutch for "Let's All Run"—and Ame was serious about the "All" part. The baseball guys had gotten into hot water because it looked like they were breaking away from the community, wearing strange clothes and spending the weekends in "English" ballparks, so Ame wasn't going to repeat that mistake. Vella Shpringa adopted a motto, "The joy of running in community," which could just as well have been "Trust us, this ain't baseball." Rather than keeping those Sunday-afternoon runs secret, Ame reached out to other youth groups. Jake Beiler showed up because he and a friend wanted to get in shape for a coming-of-age expedition to hike the Grand Canyon. Lilian found it was a fun way to spend evenings with her boyfriend, Ben. Ame even invited Ivan, a married friend with children, because he knew Ivan liked to walk his two youngest in the stroller before bedtime. *Come jog them to sleep!*

Bit by bit, Vella Shpringa began to grow. Still, the future of Amish running remained in doubt until Terry Yoder—the notorious Naked Mennonite himself—had a stroke of brilliance.

One sweltering summer day, Terry realized that a full moon was coming. Instead of suffering through their weekly long run under the sun, he suggested to Jim Smucker, why not go by moonlight? One thing that makes Amish country so special, he pointed out, is the gorgeous night sky. Out there in the "Valley of No Wires," as the literally wireless countryside around Bird-in-Hand is known, there are no streetlights, no phone poles, no backyard LEDs, nothing but a soft black landscape and the glorious, starry light show overhead. Jim agreed to give it a try. He and Terry had such a magical night, so breathtaking and serene, that they couldn't wait for the next lunar cycle. They urged Ame and his gang to join them, and it wasn't long before the Full Moon Runs evolved into a mobile monthly party.

The first time I was invited, I arrived at Ivan's farm just before sunset. Vella Shpringa members take turns hosting the run, shifting each month from one farm to another. The hosts map out two distances, usually about five and ten miles, and enlist their parents, spouses, kids, and neighbors to lay out a whomper of a post-run picnic while the runners are on the road. When I got to Ivan's, thirty or so runners were on the front lawn, a mixed crew of men and women, mostly Amish and Mennonite. Everyone mingled and stretched, while veteran Vella Shpringa-ers welcomed us newcomers and partnered up to guide us so we wouldn't go astray in the dark.

As the moon rose, we set off. The Naked Mennonite had his shirt off before we were even out of the driveway, but no one seemed to mind. No headlamps were needed; our eyes gradually adjusted as daylight faded, until all I could see were stars, the glow of oil lamps in farmhouse windows, and the dark silhouette of ancient barns. We padded through the silence, occasionally chatting but mostly quiet, content to enjoy the sounds of the night and the satisfying rasp of our own laboring lungs.

Suddenly, flashing lights lit up the road ahead. It looked like the scene of an accident, though I hadn't heard a thing and couldn't see any wreckage. "Our friends came to help," explained Jim Smucker. "They worry about us." The Bird-in-Hand Volunteer Fire Company knew we'd be out on the dark roads tonight, so they'd dispatched a few volunteers with emergency vehicles to safeguard some of the dicier intersections. "Got cold water, anybody needs one," a firefighter offered, holding out a Poland Spring bottle. "Y'all are looking good. Except Jim."

The Bird-in-Hand fire company has always had a special bond with Vella Shpringa. For those of us in Lancaster's countryside, our neighbors are the ones who come to the rescue when we're in trouble. Instead of professional firefighters, we have volunteers like John Esh, my local grocer; and Jason Tucker, a lawn mower repairman; and Sam Esh, a young dairy farmer. They monitor citizen-band radios and drop tools whenever they hear a distress call, leaving everyday life behind to plunge into all kinds of trouble. But on October 2, 2006, Bird-in-Hand volunteers ran into something they'd never seen before—and hope they can someday forget. They were among the first responders to race to an Amish school when a local milk truck driver went on a killing rampage, murdering five girls and critically injuring five others before shooting himself.

Three years after that horrible day, Jim Smucker enlisted the Amish community to help create the Bird-in-Hand Half Marathon. By modern race standards, it's primitive. There's no big expo full of merchandise, just a registration tent in a hayfield. There's no *get-fired-up!* music, only a Mennonite family on their front porch at Mile 2 singing gospel. But Bird-in-Hand is now ranked as one of the best and most memorable races in the country, partly because of the sheer beauty of the Valley of No Wires, but mostly because of the warmth and friendliness of the Amish hosts. All the aid stations are at Amish farms, with rows of children holding out cups and chanting "*Vater, vater, vater,* Powerade, Powerade . . ." All the food for the post-race picnic—the mountains of barbecued chicken and home-baked beans, the cider doughnuts and schnitz pies—is donated and served by Amish families. Every one of the finishers' medals is a real horseshoe that's been hand-burnished and laced by a Vella Shpringa volunteer. Everyone goes home with a race tee reminding them of the Vella Shpringa motto, "The joy of running in community." And every penny that's raised is donated by the Amish community to their friends in the Bird-in-Hand Volunteer Fire Company.

But if you think communal + joyful = slow & pokey, you should have been at that Full Moon Run. I'd chosen the shorter, five-mile route, since it was my first time and I didn't know what to expect. It was a warm night and I was struggling to keep up with Jim and Lilian, who didn't seem bothered by the heat even though she was in her heavy dress, apron, and starched headcovering. When we got back to Ivan's farm, I was glad to find a table in the driveway loaded with jugs of lemonade. I gulped down a cup and was about to make a hard move on the food table when I heard a stampede behind me.

Ame and the other ten-mile runners were already storming down the driveway, finishing ten miles in barely more time than it took me to run five.

"Man, those guys are getting quick," I remarked to Jim.

"This? This is just fun. Wait till you see them race."

Everything the Amish have learned over the past three hundred years about how to rely on their own bodies and build them into high-performance machines, the Vella Shpringa gang has applied to running. And it's paying off beautifully: Ame has sliced a full hour off his marathon best, improving from 3:59 to a sizzling 2:54. Ben Zook wanted to see if a tall, muscular runner like him could break five minutes in the mile *and* three hours in a marathon; within a year, he'd nailed both. As a six-man team, the Amish runners have won *three* Ragnar Relays covering distances from 128 to 200 miles. Leroy Stoltzfus was even featured in "Ripley's Believe It or Not!" when he was spotted near the front of the pack in the Harrisburg Marathon in his long pants and suspenders, finishing in just over three hours. Several times, Liz has shown up at the starting line of 5Ks and 10Ks in her full-length dress and outrun every other woman in the field to win.

"She'd do very well in fifty-mile races," said Ame, who has run a hundred-miler himself. "She's a really good distance runner." Vella Shpringa runners once road-tripped to the Niagara Marathon because they'd heard it was a fast course, only to get battered by a bitter headwind. The guys struggled to finish, then turned around to find Liz right behind them with a grin on her face and a 3:30 personal best.

"It took me eight tries to qualify for Boston. She did it on her first," Ame marveled. Boasting is *way* non-Amish, but when it comes to Liz, Ame can't resist. Besides, war stories are half the secret of Vella Shpringa's success. That's a little hack the Amish figured out long before they got into running: Dreams are the beginning of every new adventure, and our greatest dreams come from the person right in front of us. The Amish don't watch TV or movies, or even listen to the radio, so their sole source of entertain-

Amos and Liz, fifty miles into Amos's first hundred-mile ultramarathon

ment is the tales they tell one another. No wonder Vella Shpringa has grown so quickly; nothing is more inspiring than hearing your buddy tell a story while you're thinking, Well, hell. If he can run ten miles in the dark, why can't I?

That night at the Full Moon Run, we're all free to head off as soon as we finish, but even though it's late on a work night and a bunch of these guys still have to scooter home, no one leaves until we've followed two traditions. First, everyone waits in the driveway for the last straggler to make it in. Then, we load up on food and sprawl on the grass in our sweaty clothes, looking up at the stars and trading tales as Ivan passes around more of that insanely delicious salsa he made from his own garden vegetables.

After one of these nights, Ame invited Liz and her sister, Emma, to join him and some friends for a run on a beautiful but very rocky route on the historic Conestoga Trail. It was a rainy Sunday, making the slick stones treacherous, and barely a mile into the woods, Ame suddenly flew off the steep embankment and crashed down into the creek. The other runners were scrambling down to help him when he emerged with something in his hand: a heart-shaped rock. He held it out to Liz, and asked her to marry him.

Liz said yes, but Ame being Ame, he wanted to finish the rest of the six miles before going home to tell their families. "After that, she was so overjoyed I could barely keep up with her," Ame told me. "Fastest run I've ever had with her."

On their wedding day, Liz and Ame met at three thirty in the morning for a six-mile run. For Ame, it was a chance to reflect on just how weird his life had become. Normal would have meant ghosting off on his own to indulge this taboo little fascination of his, the way Jim Smucker's uncle did, before eventually giving it up and returning to Amish tradition. But Ame took a chance, and instead of hiding a shameful little secret of his own, he created something big and joyful for everyone else. He found the true Amish-ness at the heart of running, and his own running took off.

"For us, it's very unusual to do things alone," Liz once told me. "We're used to working together and having our fun as a group. That's the only reason I began running in the first place. I enjoyed those afternoons with my friends."

But early that morning it was just the two of them, side by side in the dark, before the sun came up and the buggies began to arrive. Then it was time for Liz and Ame to rejoin their families, their friends, and fellow runners, and begin their next adventure.

14

Matildonkey

"What do you think?" I asked Tanya. "Bad idea?"

"Maybe," she said, as she struggled to settle Flower. "This could be a *looong* night."

In early November, two months after Sherman arrived, I volunteered to host the Full Moon Run. We'd brought the donkeys out of their pasture in advance and tied them to the gate beside the driveway, but as the minivans pulled in and Amish runners piled out, phantom-like in the dark in their black pants and jackets, Sherman and Flower went on red alert, their ears shooting straight up like they were being robbed at gunpoint. *Too weird*, they decided, and began squirming, trying to bust free of their halters.

"Put them back?" I asked.

"Ah, let's go for it," Tanya said. "It'll either work great, or fail fast."

I'd wanted to try donkey running with the Vella Shpringa crew because I was hoping they could solve a structural design flaw we'd encountered with Tanya: we couldn't figure out a way to split her in two. During our training sessions with Sherman, Tanya and

Flower needed to ride up in front so Sherman would follow. At the same time, I needed Tanya to bring up the rear so she could correct my mistakes. I was still a rookie with this ground-driving business, and every once in a while, I'd find myself tangled in the rope and turned around backward with no idea what the hell just happened.

But the Amish guys have been training horses since they were kids, and Vella Shpringa brought another rare skill to the table: because they're such strong runners, they might be able to work with Sherman on the open road, trotting along behind him instead of walking in circles in a corral. Just by luck, I might be living next to an undiscovered talent pool of expert burro racers: where else are you going to find master horsemen with Boston Marathon speed? It was a good thought, undermined only by the dumbassery of forgetting that donkey training may not combine too well with darkness. We'd never taken Sherman and Flower out at night before, and when headlights dazzled their eyes and they were suddenly surrounded by shadowy strangers, they began to freak.

"Is that the famous Sherman?" someone called.

A van door slammed, and out of the darkness came Jake Beiler, one of Vella Shpringa's unofficial group leaders. Jake is tall and slender, but strong as a grizzly; one year at the finish of the Bird-in-Hand Half Marathon, Jake nearly single-handedly upended me and dunked me headfirst in the water-bottle barrel to cool off. Jake saw the donkeys were fretting and took command. He quickly switched off his headlamp and kept his hands low, approaching slowly. He moved his head around until he caught Sherman's eye, locking gazes to let Sherman know he wasn't in any danger.

"So this is our new friend," Jake said, his voice low and reassuring. Sherman eyed him warily but held still when Jake stroked his head and scratched him under the jaw. Around us, vans and pickup trucks continued to arrive, filling the driveway and squeezing into rows across the lawn in front of the house. The murmur of voices grew louder, a stew of English and Pennsylvania Dutch, as runners who hadn't seen one another since the last full moon traded greetings and loosened up.

Jake gave Sherman's noggin one last scruffing, then went to call everyone together. "If we're doing this, let's go now," Tanya said. "We're going to need a head start." Sherman and Flower had given up trying to escape and begun trying to hide behind each other, circling around nose to butt and tangling their ropes into spaghetti. The only way to calm them, Tanya thought, would be to get out ahead of the group and see if we could coax the donkeys into a running groove before everyone else caught up with us.

"Okay," I told Tanya. "Let's see how far we get."

I quickly explained our plan to Jake and my superhuman mutant ninja friend, Steve, a seventy-one-year-old retired watchmaker who's so indestructible that he once joined me on a seven-mile trail run even though his arm was in a sling from a freshly broken collarbone. The day before, Steve and I had taken a big sack of flour and laid out two routes along the backroads, chucking out handfuls to mark directions for five or eight miles. Under the moonlight, the flour should be visible enough for everyone to follow the turns without me.

"Give us about a twenty-minute lead, then turn everyone loose," I told Steve and Jake.

I unknotted Sherman's rope and yanked it free from the gate. I turned to make sure Flower and Tanya were set, while Sherman began trotting down the driveway—

And kept on trotting. As soon as he realized he could outrun the commotion behind him, the Wild Thing was off. I watched him go, so impressed by his speed and initiative that it took a few beats before I realized the rope was about to jerk out of my hand as Sherman disappeared into the dark. I sprinted after the fugitive, while Tanya swung herself onto Flower and joined the pursuit. When I caught up with Sherm, he didn't seem to be escaping; he was clip-clopping happily along like a thoroughbred on parade. I bent down and grabbed the rope, but he never broke stride, cruising steadily at a crisp jog.

"What's he up to?" I asked Tanya.

"Beats me," she said. She held Flower back to see if Sherman would slow down and let his girlfriend lead, as usual, but he seemed oblivious. After about a quarter mile, we hit a long grinder of an uphill slope and Sherman didn't hesitate; he shifted into climbing gear and streamed along so smoothly, I was gasping to keep up.

"Holy crap," I panted. "What's gotten into him?"

"Anything to mess with your mind," Tanya replied. "That's donkeys. Never what you expect."

We crested the hill and kept flowing. It was a gorgeous evening, with the valley below us glowing silver in the moonlight, but Tanya's mind kept circling back to Sherman. "Nighttime probably lets him focus," Tanya speculated, keeping her voice low. We were both whispering, afraid to jolt Sherman out of whatever voodoo trance he was in. "Tunnel vision. All he's sensing is Flower and the road ahead, so he doesn't have to process all the scary stuff he sees in the daytime."

Sherman ignored us, barreling straight into the night, a donkey on a mission. Only at Mile 2 did I spot the first danger sign: Sherman's ears perked up and swiveled back, detecting some menace in the silence around us. A few moments later, a shout rang out.

"Finally!" a distant voice called, and then I heard pattering feet. The first Amish runners were closing in fast, surging into view as they topped the hill behind us. I tightened my grip on Sherman's rope, prepared to haul back if he got spooked, but other than the ear twitch, he didn't flinch.

"You guys are flying," Jake said, pulling alongside us. Beside him was Laura Kline, the 2012 World Champion duathlete and U.S. National Team triathlete. As many times as I've seen Laura on these runs, it still came as a jolt whenever I'd spot a gang of young Amish men and find, in the middle, this ripped elite athlete in her sleek compression gear emblazoned with sponsor logos. Laura had moved to the Southern End from Baltimore a few years ago when she heard about the spectacular running trails along the Susquehanna, and she soon became a Vella Shpringa regular. Her speed speaks for itself, but it's her O.G. work ethic that really bonds her with the Amish guys; I've watched Laura run for miles through unbroken snow up to her knees, and charge into the woods during a freezing winter storm that coated the rocks with ice. Laura had recently moved to New Paltz, New York, but when she heard that tonight's Full Moon Run might feature donkeys, she had to make the four-hour drive back down to the Southern End to check it out.

"Looks like Sherman's got some go in him," Jake said. "Mind if I take a try?"

I opened my mouth to explain why that was a bad idea, then shut it and handed him the rope. I hated to tamper with Sherman's sudden miraculous mojo, but the whole point of bringing the donkeys out tonight was to see what I could learn from Vella Shpringa. Jake may not have run with a donkey before, but I had to believe his lifetime of animal savvy would let him suss things out. Sure enough, Jake expertly coiled the rope in his left hand and, with the right, gave Sherman a reassuring pat on the rump. The rest of the crew formed a flying wedge with Laura setting the pace, surrounding Sherm so closely on all sides that all I could see was two long ears jutting up from a circle of bobbing heads.

"Get up there, fella," Jake said as we rounded a curve in the road and approached a long downhill. Sherman was already at a brisk trot, but at Jake's command, he accelerated into a canter. I dropped off the pace a little so I could get a better look at Jake's technique. He was only a few inches from Sherman's left haunch, keeping himself much closer than I usually did. Every few strides, Jake clucked his tongue or gave Sherm a little pat with his hand, gently reminding him that they were on the job. But Sherman showed no sign of slowing, even when the rope switched hands from Jake to Jonathan to Elam. Everyone was eager to take a turn, and they all handled Sherman with the same confidence and purpose. I'm not even sure Sherm was aware when a new runner stepped in.

As we breezed through Mile 3, eight hooves and twelve feet were pattering in unison, a single drumbeat uniting the tribe. I loved the way everyone instinctively synced their pace, adjusting their speed up or down a notch to make sure that humans and animals were all flowing comfortably. We were having such a blast, it took a good half mile before the distress signals from my legs and lungs made their way to my brain and I realized I was in trouble. Sherman and Flower were keeping up beautifully, but for me, the party was coming to an end in a matter of seconds. Even downhill, I was out of my depth with this crew, and we were now at the base of another uphill slog. No way was I tackling that beast alongside Laura Kline.

"I'm out," I said, slowing down and peeling off from the group. Sherman and Flower could keep going, I figured, and Tanya could wait for me with Sherman at the top of the hill. Or something; going too fast was never a problem I thought I'd have to plan for. But when I dropped back, Sherman suddenly balked and U-turned, scattering the runners like bowling pins and T-boning so sharply in front of Jake that he nearly flipped over Sherman's back.

Jake caught his balance and handed me back the rope. "Yeah, he's had enough of me," he said.

"'Bye, cuties," Laura called, rubbing Flower's muzzle. "See you back at the ranch."

With that, she and the Amish guys stormed the hill and were soon out of sight. I took a sec to catch my wind, then Tanya and I started out on our own. But the sorcerer's spell was broken: when we tried to get Sherman and Flower running again, they suddenly remembered they were donkeys. They began futzing around, veering onto the grass for snacks and taking playful nips at each other. Bit by bit, we got them up the hill, but by the time we reached the top, an evening mist had splattered the road ahead with dozens of Flower phobia triggers: damp patches. Flower minced her way to the bottom, one doubtful step at a time.

"Fun while it lasted," Tanya said.

"Unreal," I agreed. We'd run little more than half the short course, but we'd run it like real burro racers. "Sherman was *scorching*."

"Let's end happy and walk them in," Tanya said. "It's all about building a bank of goodwill. You never want to draw down too much. You want to keep adding to the reserve, one good experience after another. Someday, you're going to ask Sherman to do something he doesn't like, and because you've built up the bank, he's going to surprise you."

Sherman meandered nonchalantly as we hiked the two miles home, snatching bites of roadside weeds and paying no attention to the intense postgame analysis Tanya and I were conducting about what the hell just happened. Somehow we'd mixed a magic potion, but we couldn't figure out how. Was it the Amish guys? They were great, but Sherm was already on a roll before they caught up. The night running? Maybe, but the dark didn't stop him from futzing at the beginning or quitting in the middle. Did we blow it by pushing too hard? *He* wasn't the one who punked first, Tanya reminded me; that was me. Besides, Sherm bounced right back when it came to messing around with Flower.

But for a moment, something had definitely clicked. We still hadn't figured it out by the time we got back to the house, where some of the Vella Shpringa guys and seventy-one-year-old Steve had already arrived and taken over my hosting duties. Burgers were sizzling in the firepit, Steve was ladling out the vegan chili with Guinness Stout that I'd made in the Crock-Pot in advance for vegetarian Tanya and no-meat-athlete Laura, and I suddenly realized I was ravenous. I loaded a plate and joined the gang around the fire, and as I got caught up in the jokes and storytelling, I forgot about solving the mystery of Sherman's miracle vanishing makeover.

But Tanya didn't. By the time she went to bed, she had her answer.

The next morning, the blare of a car horn announced that Tanya had something to share. Her head popped through the driver's window as she pulled into our driveway in her van. "Your secret weapon has arrived!" she shouted. She climbed out and stood by the rear door, her hand poised dramatically over the handle, as I pulled on muck boots and came outside.

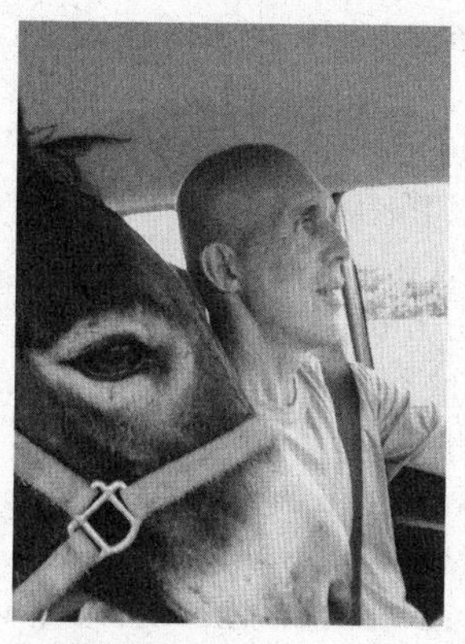

"Matildonkey" off for a hoof-trimming with me in the minivan

"Behold!" she said, throwing open the Dodge's back door. There, in all its tiny glory, was the mini donkey she'd rescued from the slaughterhouse.

"Matilda!" I said. "I thought you adopted her out to some family."

"Yeah, she only lasted there a day. They had a dog that came at her. Matilda kicked it so hard its leg had to be amputated."

"How'd you get her in the car?" From what I'd seen of burro racing, every story seemed to begin, end, or revolve around a donkey going full Hulk Smash to avoid being loaded into a trailer. And here was Matilda, a proven bonebreaker, chilling in the back of an old Dodge van like she was going out for ice cream.

"She's up for anything," Tanya said. "Watch this." She gave the halter rope a gentle pull and crooned, "C'mere, 'Tilda. Come on, 'Tildonkey." Matilda squirmed her way through the narrow gap between the front and back seats, took a look out the door, and hopped down. She paused for Tanya to rub her ears in appreciation, then sauntered over to the fence gate where Sherman had been tracking her entrance with rapt attention. We watched them get acquainted while Tanya told me about her brainstorm.

After the Full Moon Run, she'd gone home still itching with curiosity about what had inspired Sherman. She kept chewing over whether the secret ingredient was the moonlight, or the hectic driveway scene, or some other weirdness we hadn't noticed, until it finally hit her that the answer was "Yes." Yes, it was the spooky darkness, and the hubbub, and the strange German voices in the dark, and everything else that combined to make the night a donkey-running thrillfest. We'd thrown a bunch of scary new stuff at Sherman all at once, and scary new stuff was exactly what Sherman wanted.

"We're teaching Sherman his job is running, right?" Tanya said. "But his job doesn't have to be boring." Until last night, we'd always stuck to the same gravel road through the woods whenever we took Sherman out. We wanted to build his strength and confidence without overwhelming him, so we'd kept him close to home and as far from commotion as possible. But we forgot one thing: Donkeys love a walk on the wild side. Out on their own, they're always on the move, constantly roving, finding places they've never seen before in search of food that only they can reach. Their survival depends on long-range roving, which is why they've developed such extraordinary endurance, sure-footedness, and risk assessment. Sherman was born with the badlands in his blood, and after six weeks of gentle progress, we weren't protecting him anymore; we were boring the shit out of him.

"No wonder he ran like a champ last night," Tanya said. "He was partying." His first impulse was to bolt back to the safety of his little barn, but once he connected with Jake and realized some kind of catch-and-pursue game might be afoot—an up-tempo version of his old Donkey Tao days with Chili Dog—he couldn't wait to get started. My plan for the Amish runners had worked, except not in the way I'd expected; I wanted Vella Shpringa to teach me about donkeys, but they taught the donkeys something about Vella Shpringa. Sherman had discovered the joy of running in community, and he loved it. So much, in fact, that when Laura and the Amish gang left us behind, as far as Sherman was concerned the adventure was over. Why keep playing when all your playmates have gone home?

Sherm needs a play buddy that will never quit, Tanya thought. He needs Matilda.

"She's the piece to the puzzle that we've been missing," Tanya explained. Sherman and Flower are ready to explore, but when things get a little spooky, they stick so close to each other they end up walking in circles. But not Matilda. Whatever Matilda had survived in her past, it had made her Sherman's opposite: where he's calculating and guarded, she's curious and fearless. And despite her short legs, Matilda is also a solid runner; every once in a while, Tanya would take Matilda for jaunts alongside her horse carriage, and Matilda kept up fine.

"She's a little badass," Tanya said.

"So what's the plan?" I asked. "Leave Flower and you run Matilda?"

"Oh, good God, I'd die." Tanya couldn't remember the last time she ran anywhere without a horse under her. If Matilda was going to join this team, this team would need another recruit. My wife, Mika, was home, but she's a passionate African and Hawaiian dancer who never really understood why anyone would voluntarily spend an hour of their life repeating the same movement over and over in a straight line without at least a bare-chested drummer along to liven things up. Runners like to boast that "our sport is your sport's punishment," and Mika couldn't agree more. But she'd do just about anything for Sherman, so I went inside to make my pitch.

Five minutes later, Mika was tying on sneakers. "What should I do?" she asked.

"Just hold on to the rope," Tanya replied. "Matilda will take care of the rest."

Tanya heaved her saddle onto Flower and walked her out. Mika and I followed with Sherman and Matilda, but Matilda wasn't following anybody. She jerked in front and took position at the head of our little herd, swishing her tail menacingly at Flower to make it clear who was in charge. Sherman was already starry-eyed over this sassy little alpha girl and hurried to squeeze in beside her, bumping Flower ahead and causing Matilda to leapfrog forward again. Tanya hadn't even given a command and already we were prancing down the road.

"Ready for some fun?" Tanya asked. "Looks like they are."

Tanya clucked to Flower and we were off, with little Matilda jogging stubbornly in front and the other two donkeys tight behind. In no time, we were up the short hill to the gravel road and approaching the creek culvert where Flower always observed her morning ritual of a minor meltdown. It was a good opportunity to slow down and catch my wind, but Field Marshal Matilda never gave me a chance. She led her troops straight across, and for the first time, Flower never gave the creek a second look.

"'Tildonkey! Good girl!" Tanya cheered. "How're you doing, Mika?"

"She's amazing," Mika panted, likewise struggling to handle Matilda's no-warm-up work ethic. "But I could use a quick break."

"Yeah, I don't know if I can handle this much fun," I agreed.

We reined everyone in and gave them a beat to munch grass while we sucked air. We kept it super short because we didn't want to spoil the flow, but the flow turned out to be unspoilable. As soon as Tanya nudged Flower back into a walk, Matilda shot off to get in front, determined to keep her position at the point of the spear. We flew down the gravel road toward the waterfall, and once again, Flower 2.0 streamed past this personal nightmare so quickly, I was robbed of my usual time-out.

Fine by me, though. The farther Matilda led her little brigade, the more Mika and Tanya and I craned around to catch one another's eyes and exchange baffled grins. There was a feeling in the air, a sense of pre-game excitement that we were on the verge of an extraordinary breakthrough. Tanya was onto something, and even the donkeys knew it.

15

Gang of Three

Everything okay? Mika texted to Tanya.

No reply.

Over Thanksgiving, we had all taken some downtime while Tanya was away with her parents. We were ready to hit it the day Tanya was due home, but early that morning she messaged to cancel. She didn't give a reason, which was fine, but said nothing about rescheduling, which was strange.

Mika tried her that evening, and again the next morning, but there was still no word. By noon we were getting worried, so I drove over the hill to Christmas Wish Farm. The house is nearly invisible from the road, a small, Hobbity cabin tucked away in the woods behind an Amish farm and nearly swallowed by a forest that slopes down to the Susquehanna River. Every time we go, a stampede of barking dogs greets us before we're even halfway down the gravel lane, and then Tanya and Scott will come banging through the screen door with a smile and a wave. This time, only the dogs showed up.

Really weird, I thought. I'd never had to actually knock on their door before.

I rapped, then harder. Inside, more dogs barked. If Tanya and Scott were away, wouldn't all the dogs be either inside or out? Something wasn't right. I hammered the door again, which made the dogs inside frantic. That's it, I'm going in, I decided—then remembered that a few of the indoor dogs are Dobermans. Maybe better knock some more. I pounded with the side of my fist, calling Scott and Tanya's names, until I finally heard someone shushing the dogs.

The door cracked open, and behind it was a Tanya I'd never seen before. She looked disoriented and utterly drained, as if she'd just fallen asleep after being awake for days. Which, unfortunately, she had.

"Scott's gone," Tanya said. "Totally done and gone." Tanya had been out of town with her parents over Thanksgiving, and when she got back, Scott hit her with a thunderbolt: he was packed and leaving. Tanya was completely blindsided. Until that moment, she'd thought their life together was wonderful and special. Not only were they perfect for each other, they were perfect for no one else. They were a rare breed of geek, with an equal love of horror films and horsemanship. Where are you going to find another soul mate who's just as content as you are to drive Victorian-style carriages by day and watch *Evil Dead II* by night? Only afterward, during those long, miserable nights that Tanya spent trying to understand where things had gone wrong, did the Sixth Sense pattern of clues pop out at her: Scott's eagerness to do chores at home rather than join her at horse shows, his sudden interest in running, although only at work and only during lunch with a coworker who was "just a friend" . . .

Now, suddenly, everything was collapsing. Tanya couldn't live alone on the farm, caring for the animals and handling the back-breaking daily chores and earning enough to pay all the bills by herself—could she? But if not, what would happen to Flower, and Matilda, and their dogs and horses? What would happen to *her*? Tanya realized there was no way she could run the farm on her own . . . just before she resolved that there was no way in hell she'd give it up. Whatever it took, however hard she had to work, she was going to figure this out.

Terrifying as it was, at least reaching the decision allowed her to relax and close her eyes for the first time in days—which, of course, was when I came thumping on the door. Already that morning, Tanya had gone to the neighboring Amish farms to let them know that anytime they needed a driver to go anywhere, at any hour, she was for hire. She wasn't going to abandon Sherman, Tanya promised me, but for now, she needed some time to sort out her finances and ramp up her paying jobs.

"Absolutely," I assured her. "You worry about you. We'll be fine until you're ready."

"You better," Tanya said, with the don't-piss-me-off tone she uses when Flower is backing away from a Burger King wrapper on the side of the road. "Because I'm going to Colorado." Tanya knew she was about to begin the fight of her life to keep her farm afloat. She needed something to look forward to, and right now the only bright spot in the future was the hope that by next summer, she and Sherman would both be so strong that the only thing they'd have to worry about would be getting to the starting line on time for a race across the Rocky Mountains.

We're toast, I realized on the drive home. Totally torpedoed.

I was all gung-ho in front of Tanya, but even as I was reassuring her that nah, c'mon, we'll be just dandy, I could feel the dread rising as a small part of my mind was already calculating the wreckage. We were facing crazy odds and a tight deadline, and now the three most important members of Sherman's team—Tanya, Scott, and Flower: his coach, medic, and personal trainer—were out.

No Scott meant no one to tend to Sherman's healing but still misshapen hooves. No Tanya meant we'd lost both a donkey whisperer *and* a donkey, because what would we do with Flower? We'd made our biggest breakthrough when we decided to follow Vella Shpringa's lead and surround Sherman with a band of friends, but without Tanya, we now had more donkeys than donkey handlers.

Sherman and his new buddies had already bonded into such a Gang of Three that I cringed at the thought of taking two of them out for a run and leaving the third behind.

"Let's see how it goes," I told Mika, after I'd gotten home and filled her in. There was no telling if Tanya would be back in weeks, months, or ever, so we might as well find out now how hard it would be to go it alone. We headed outside, and as soon as the donkeys heard the chain rattle on the gate, they came frisking toward us.

"Look at them, ready to play," Mika said. "That's a good sign."

We haltered and roped Matilda and Sherman, but I didn't know how we'd get them outside the fence without Flower stampeding along behind them. "Turn her around a sec," Mika said. I led Flower in a slow circle, and before we'd gone the full 360, Mika somehow funneled Matilda and Sherman through the gate and slid out behind them. I squeezed through, chaining the gate securely. Amazingly, Flower just stood and watched.

"We'd better haul ass while we can," I said. Mika clucked to Matilda and we were off, trotting down the driveway. Behind me, Flower began to huff nervously. I didn't dare look back, thinking that maybe if I didn't make eye contact and got Sherman moving, Flower would just chill out and go on with her day. But Flower's snorts kept getting louder and faster as she cycled from confusion, to concern, to—

"Oh my god!" Mika said, wheeling around.

—Four-alarm panic.

Flower had erupted with an earsplitting blast of utter despair, a wail both deafening and heartbreaking, the sob of the world's saddest car alarm. If you've ever heard a donkey bray, you never again have to wonder what the souls of the damned twisting on the pitchforks of eternal torment sound like. Personally, I think Flower was hamming it up a little; I know for a fact that when Mika and Matilda did their Talking Donkey trick, Matilda could bellow out a pity party on command.*

* Yes, we've got video.

But Flower's misery was convincing enough for Matilda and Sherman. They wrenched themselves around while Flower was still in mid-bleat and strained to hurry back. Mika and I held our ground, pulling against the ropes as we debated our next move. We were right where we'd started two months ago, stuck in the driveway with no idea what to do. I didn't know whether we should force Sherman and Matilda to come along, or respect their bond with Flower and give up on running without her. Were we good parents who knew our little ones would love kindergarten once we dropped them off, or Tiger Moms who didn't know when to ease up?

"Sherman is going to have to run by himself on race day," I reasoned, which sounded stupid and unlikely even as it was coming out of my mouth. On race day? We couldn't even walk fifty feet without Tanya. Why pretend we still had a prayer of going thirty miles?

Mika must have read the surrender in my face and decided to take charge. "If Matilda goes, Sherman might follow," she said. "And anything we start, we have to finish. Right?"

Mika braced the rope across her hips and marched toward the street, abandoning the ground-driving technique we'd been working on and reverting to the old-school tug-of-war style, pulling Matilda along from in front one grudging step at a time. I followed Mika's lead, hauling Sherman down the driveway with me. Flower was storming back and forth along the fence line, bellowing for us to come back and get her, but Mika and I kept marching the two donkeys up the street to the turnoff onto the gravel road.

As we made the turn and disappeared behind the trees, Flower's braying began to fade in the distance. Sherman and Matilda's resistance eased as well, and with a little intensive clucking and encouragement, we got them trotting—for a while. Every few dozen yards, one or both would suddenly have second thoughts and pivot toward home, forcing Mika and me to constantly be on the lookout, scoping their body language for early-warning signs so we could scamper from right flank to left to cut off whatever U-turns were brewing between those furry ears.

By the time we got to the end of the gravel road, Mika and I had packed about three miles of running into a one-mile stretch, sprinting in zigzags to keep the veering donkeys moving forward. Now I really understood what Tanya brought to the game—not only her expertise, but her constant cowgirl maneuvering in the saddle as she deftly wheeled Flower around to stay one step ahead of any Sherman dumbfuckery. Without her, Mika and I could never relax into a groove; we were ranch hands more than runners, playing constant defense in a full-court press against breakaway critters. After one mile, we were fried.

"Bring 'em home?" I asked.

"I'm done," Mika agreed.

I started to pull Sherman away from the grass he was chomping, but I didn't have to. As soon as I walked toward him, he lifted his head and started walking up the gravel road. "Yup," I said. "Flower time." The closer I got, the more he picked up the pace, until we both shifted naturally from a walk to a jog. Matilda, who was a little more deeply invested in her munching, glanced up in surprise to find Sherman forty yards gone and opening ground. She closed the gap in no time, and together, the two little Seabiscuits raced toward home.

Flower was browsing quietly when we came around the turn toward the house, her separation anxiety apparently eased by some tasty wild greens she was pulling out from under the fence. But her head jerked up when she heard hoofbeats drumming on the pavement. She scampered back to the gate, letting loose a roof-shaking bray. Matilda and Sherman hollered back, whooping it up without missing a step as we charged down the road to the gate. Getting them inside, however, was madness; as soon as I lifted the chain, Flower and Matilda collided head-on, with Flower bulling her way out as Matilda was shoving her way in. Sherman, meanwhile, was dancing around in the middle, so thrilled by all the attention and companionship that he was happy to stay where he was, twisting and squirming while Mika and I dodged hooves and untangled ropes to sort everyone out.

I wasn't looking forward to repeating this rodeo the next morning. The only sensible move was to bring Flower back to Tanya, but for at least three reasons that was too heartless to suggest: Sherman was devoted to Flower and Matilda; Flower was devoted to Matilda and Sherman; and Mika and I were devoted to Tanya. We couldn't bust up the Gang of Three and saddle Tanya with another mouth to feed just because we'd had a rocky time during our first run on our own. All we had to do was be patient and consistent, and the donkeys would adapt. We knew that; we'd seen it happen, over and over, since the day Sherman arrived.

"We're going to have to run them every day, no matter what," I told Mika.

"First thing every morning," she agreed. "Get the girls to school, then run."

"First thing," I promised.

First thing the next morning, I discovered we were out of chicken feed. Before messing with any donkeys, I had to hustle to the feed mill for half a dozen fifty-pound sacks and carry them up to the shed. I also picked up a salt block for the donkeys and a few bags of sheep feed, which required shoulder-carrying to a different shed. By the time everything was stored and everyone was fed—cats, chickens, ducks, geese, sheep, goats, and donkeys—I was famished myself. I went in for breakfast, but found the fireplace and wood-pellet stove were burning low, so it was right back out again, this time for a fifty-pound bag of pellets and an armload of split locust logs.

One thing you discover on even a small farm is that come winter, everything weighs fifty pounds and you're always carrying it someplace: bales of hay to the feeders, bales of straw to the stalls, five-gallon buckets from the creek to the frozen water trough, chunks of logs from the woods for splitting, split firewood into the house for heating, and those eight-foot fence posts still piled in the yard, which, I suddenly remembered, absolutely had to be hand-dug and pounded in place *right away* before the ground froze.

I didn't want to run as soon as I finished eating, anyway, so after breakfast, I decided to work on the fence posts for a while. A bunch of e-mails and a writing assignment also needed attention, and then Mika suggested lunch. By three that afternoon, the December sun was getting low and we remembered one of the girls had a basketball game . . .

All of which is to say, we never did get the donkeys out that day. Or the next; a winter storm was threatening, so we needed to get 150 bales of hay delivered from our neighbor and then haul each bale across the pasture and stack it in the bays in case we got socked by an early snow. All the other animals trailed along to snatch bites off the bales while we were carrying hay except the donkeys; they were wise to us now and kept their distance in case we were planning to abduct two of them again. By afternoon, we could see that the storm wouldn't amount to more than a chilly drizzle, but the choice between chasing Sherman and Matilda for half an hour and watching the rain over a steaming afternoon coffee was no choice at all.

That was when we knew it was over. The days were shorter and colder and, with Christmas on the way, things were only getting busier. I knew the donkeys were only being a little fussy, and that once they got used to heading off without Flower, they'd be back to having fun again. But the hassle of chasing them in the cold, of lunging foolishly for them while our hands froze on the damp ropes and our feet got numb from the icy pasture, made it easy to keep postponing their training. When winter socked in and the roads were covered in snow, would we even try anymore?

Sherman's chances of becoming an athlete—a donkey with purpose—were slipping away. And then the phone rang, and by the time I hung up, everything had changed. A friend was in serious trouble, and the only way to help was to get Sherman back out on the roads.

16

Sick, Stressed, or House Arrest

Alarm bells went off as soon as I got Andrea Cook's message. She was brief and upbeat, but I could tell right away that something was wrong:

> *Hey, Zeke is home from school and wants to talk to Chris about running. If he's got a sec, could he give him a call?*

We've known the Cooks almost as long as we've lived in the Southern End. Our kids went to the same elementary school, where Andrea was both the beloved nurse and the human dynamo who miraculously persuaded every class to run long, sweaty laps around the ballfield every year as a charity fund-raiser. Besides her nursing, charitable activism, graduate school night courses, homemaking, and shuttling her kids twice a day to swim practice, Andrea was also training for a triathlon (of course), so occasionally we'd go on long bike rides together during her ninety minutes of "free" time in the afternoon. I liked her a lot; she and her husband, Andy, and their three kids were so genuinely kind, bright, and neighborly that you could almost forgive them all for being so damn good-looking. Zeke, in particular, won my lifelong affection when he was in fourth grade and never blamed me for giving him poison ivy when I hauled him up into a tree by a rope during my daughter's birthday party treasure hunt.

We saw a lot less of the Cooks as our kids got older and everyone's lives got more hectic. Andrea and I hadn't ridden together in a few years when I heard vague mentions that Zeke and his older sister, Ashling, hit a rough patch toward the end of high school. I didn't hear many details, only something about depression, and when I saw them next, at Ashling's graduation party, they both looked tan and strong and happy, a pair of teenage sun gods digging into nacho dip and butt-bombing cannonballs into the backyard pool. There was no hint of any lingering problems, and they both finished with stellar grades and went off to Penn State. But now, suddenly, super-student Zeke was home in the middle of the school year. And he wanted to chat with his mom's fifty-two-year-old friend about jogging? Something wasn't right.

Before talking to Zeke, I called Andrea to find out what was going on. To my surprise, Zeke picked up. I glanced at the phone: I'd accidentally dialed the home number instead of her cell. "Zeke, hey!" I stammered, groping for something to say that wouldn't put him on the spot. "I hear you're, uh . . . back for a while?"

"Yeah, I'm taking a semester at home," Zeke said. He didn't say why, which told me I shouldn't ask. He'd be commuting to one class at the University of Delaware, he said, and since that left him plenty of free time every day, one of his goals was to get back in shape. He'd been buried in lab research last semester at Penn State, and needed more sweat and sunlight in his life.

"Why don't you come over tomorrow for a run?" I offered. On the edge of my mind, an idea was taking shape. Already I could see it had some serious drawbacks, a big one being that I still had no clue what was wrong with Zeke. But if he was that same kid who'd let me hoist him fifteen feet into the air by an old clothesline, it might work. "You're up for anything, right?" I asked.

. . .

The next morning, both Zeke and Andrea showed up. Andrea had been sidelined for a while with lower-back trouble, so I hadn't expected her to come for the run. "I won't get in the way," she promised. "But I have to see what you're getting my son into."

We headed inside for a quick coffee. On the back porch, Zeke scooped up one of our mangy adopted cats and cradled it in his arms. Few visitors ever have the urge to put their hands anywhere near this battle-scarred little beast, but Zeke didn't hesitate. What a tender guy, I thought, until I realized Zeke was just lifting it for closer inspection.

"Cool," Zeke said. "Polydactyl."

"No, that's Cheetah," I corrected him, a split-second before I remembered Hemingway's Key West cats and missed a chance to sound smart. "That means six-toed, right?"

"Yup," Zeke said. "Such an interesting mutation."

Look who's talking. What other Penn State sophomore even notices a snoozing cat, let alone classifies it by proper Greek taxonomy? I'd always heard that Zeke was smart, but he's so muscular and energetic that whenever people told me that, I assumed they were too polite to add "for a jock." It dawned on me that I'd only really known Zeke as a kid. I hadn't been around him much over the past few years, so I didn't know what kind of young man he'd become, or what kind of trouble had brought him home. When I came up with my idea the night before, I was building it around courageous, courteous, thirteen-year-old Zeke, not this twenty-year-old stranger. What if he'd turned into a smart-ass slacker? If so, this experiment wasn't going to last much longer than the coffee.

"So here's the deal," I began. "Sophie finagled us into rescuing one donkey, and now we've got three." I told them all about Matilda and Flower, and Tanya's philosophy that donkeys need a purpose, and our current predicament: We couldn't run Sherman and Matilda without Flower, but Flower no longer had a rider.

If we had any hope of giving Sherman a crack at the pack burro championship this summer, we needed to get him back on the trails right away. I'd thought of asking the Amish guys for help, but they worked long days, lived too far away, and didn't have cars.

"So if you're okay with it—" I said to Zeke.

Andrea's eyes widened a little, as if she were bursting to speak but biting her tongue.

"If *you're* okay with it," I went on, turning toward her, "I'm looking for an extra runner to see if we can take Flower out on foot."

I'd never tried running with Flower, for obvious reasons. She's so strong and jumpy, all it would take was one spooky shadow and she'd be gone, ripping the rope out of my hands and hightailing it for the hills. Still, I couldn't forget that time I'd tried burro racing in Colorado and lined up next to Barb Dolan. Barb had a monster of a donkey named Dakota who was even bigger than Flower. The crowd was making Dakota nervous, but even though Barb is half my size, she handled that brute like a dance partner. She didn't just keep Dakota under control; together they were so fast, they could beat most any other racer, man or woman, over any distance.

I'm no Barb Dolan, but I was hoping that what I lacked in donkey-wrangling skills I could make up for with Matilda. There'd be no stopping Flower if she decided to tear loose, but maybe she'd just follow Sherman's lead and quietly fall into formation behind the little brown crew boss. And if that didn't work . . . well, I was pretty sure she wouldn't flee too far before returning to either our house or Tanya's. I didn't look forward to knocking on Tanya's door to ask, Hey, about that favorite donkey you lent us—seen it around? But that was a worry for later. Right now, my problem was whether Zeke—and his mom—were okay with this scheme.

"So they run next to you?" Zeke asked. "Like a dog?"

"Pretty much like a dog," I said, mostly for Mama Andrea's sake. "Like a dog that kicks. But today, we're just going to take it easy and see how Flower does. Andrea, you want to walk in front? If you lead, they'll probably follow."

"Fun!" Andrea said. "I'm in."

I wasn't sure how to divvy up the donkeys, until I realized we had no choice. I had to take Flower, since if anyone was going to get dragged around or booted, it should be me. That meant Mika would be out front with Matilda to keep things moving, leaving Zeke with the Wild Thing. I hated inflicting Sherman on any first-timer, least of all someone who was dealing with mystery issues of his own, but so be it; ya play the card ya got.

Mika doled out fistfuls of horse treats to Andrea and Zeke. The donks must have gotten over their past separation anxieties, because instead of doinking around and dodging us, today they surged right to the gate and let us slip on halters and ropes while they gobbled treats out of Zeke's and Andrea's hands.

"All right," I said. "Let's see how far we get. Andrea, take 'em out."

Andrea was so excited, she ignored both her own back pain and everything I'd said about walking and instead took off at a run. I waited behind with Flower, figuring she'd be easier to handle if Sherman and Matilda showed her what to do, but Flower had plans of her own. She was off before Matilda even had a chance to start, trotting so briskly that when Andrea glanced over her shoulder, she was looking straight into Flower's eyes. "Oh!" she exclaimed, startled. "Hello, missy."

It was kind of crazy; Flower followed Andrea so obsessively that every move Andrea made, Flower mirrored. The big donkey slowed behind Andrea on the uphill, sped up when she crossed the road, and even buttonhooked when Andrea turned around to check on the others. A few yards behind us, Matilda seemed a little miffed to find Flower out front and was steaming hard to catch up, forcing Mika to nearly sprint. Farther back, Zeke and Sherman seemed to be doing nicely. Zeke was making the rookie mistake of leading Sherman instead of driving him, but that was a simple fix. Out of everybody I somehow ended up with the easy job; all I had to do was hang on to Flower's rope and leave the heavy lifting to Andrea, who was panting hard but hanging tough.

But when we were still a good thirty yards away from the gravel road, Andrea began slowing down. "Looking good," I encouraged her, while mentally yelling *Go, go, go!* I didn't want to goad Andrea into hurting her back—not out loud, at least—but this was the worst possible place for her to stop. Flower would hit the brakes, just the way she did when the Amish gang dropped us on the Full Moon Run, and I'd be stuck on a blind curve with three unmovable animals. If Andrea couldn't run any farther, there was only one thing for me to do: it was time to try the Growl.

Barb Dolan had this thing she'd do whenever her donkey slowed down. A rumbling growl would gather in the depths of Barb's innards and slowly roll up through her throat like thunder across the prairie, ending in a sharp bark of command. Even I snapped to attention the first time I heard the Growl. Back then, my future goals for donkey running were to never, ever subject myself to that kind of misery again, so it never occurred to me to ask if the Growl was a universal donkey password or strictly a Barb-Dakota thing. I still remembered what it sounded like, though. I got as close as I could to Flower's flank, sucked in a bellyful of air, and gave it my best Barb.

"Heeeeeyyyyyyy-yup," I called.

Flower's ears didn't even twitch. I dialed up the volume and let it rip again. "Heeeyyyy—YUP!"

Andrea looked back. "Did you say stop?"

"No. 'Yup.' "

"Yup?"

"Yeah, it's this donkey thing," I hurriedly explained, but by then, it was too late. Andrea stood there, chest heaving and hands propped on her hips, listening to me babble as Flower froze to a standstill beside her. Behind us, Sherman and Matilda were already slowing to an amble, ready to crowd right on into our huddle, exactly in the kill zone of the blind curve.

"Do you mind standing over here a sec?" I asked Andrea, pointing toward a thicket of brambles between the road and the barbed-wire fence. "I'm going to try something." Without even asking why, Andrea stepped into that prickly mess. Quickly, I walked as far behind Flower as I could get, stretching her rope to the limit. Then I ran toward her, flapping my arms and giving it all the Barb I had: "hhhhheeeeeEEEEEYYYYY-YUP! hhhhheeeeeEEEEEYYYYY-YUP-YUP-YUP!"

Flower edged toward Andrea. I kept coming, barreling into the gap between them, waving my arms like a demented giant seagull. Flower tried pivoting toward Matilda and Sherman, who'd stopped a few yards away to watch this spectacle, but I cut her off. Cornered between me and Andrea, Flower had only two options: she could stampede into me and escape, or turn around and cooperate. Flower backed up two steps, giving herself running room. I gripped the rope tightly and braced as she . . .

. . . wheeled toward the gravel road. Amazing. It actually worked. I stood there, admiring my success, until the end of the rope sliding through my fingers reminded me to shake a leg. I scooted after Flower and soon caught up, positioning myself just beyond kicking range on her left flank. As we ran, I watched for any tells that she might balk while Flower side-eyed me right back. We clipped along that way, eyeing each other suspiciously as we passed AK's Saw Shop and rounded the shady bend toward the waterfall. "Damn, Flower can *move*," I thought, struggling a little to match pace as she picked up speed down the short hill.

I didn't dare look behind to see how Mika and Zeke were doing; the last thing I wanted was to remind Flower that her buddies were just a U-turn away. Flower finally skidded to a halt when we approached the Dreaded Trickling Underground Creek, which was fine by me. After that fast quarter mile, I was happy for a breather. I was also curious to see if I could repeat my luck with the Growl and get Flower going again from a standing start.

"Yeah, Flower!" Mika cheered as she and Matilda pulled up beside us. "You guys looked great."

Zeke and Sherman arrived a few moments later. "Man!" Zeke blurted. "He was impossible for a while. Then all of a sudden, he just opened it up."

"He was horrified that Flower left him," Mika said. "He couldn't even blame it on Tanya."

"I think Sherman missed the object-permanence part of his

brain development," Zeke added, which left me scratching my head until some old psych class memories bubbled back. "As soon as Flower is out of sight, he has no idea he'll ever see her again."

While we were talking, Sherman butted his big old head into Zeke's hip. Without looking down, Zeke continued to chat while absentmindedly rubbing the fur along Sherman's jaw and scruffing his mohawk mane. I was impressed. Sherman is a master of psychological torture, a true craftsman at driving a dental drill into your last nerve, and if ever there was a time for the Wild Thing to dip into his bag of tricks, feints, and all-around mindmessing, it would be when some stranger—a nervous college kid, no less—was trying to boss him around. Sherman made it clear from the start that the rope in Zeke's hand didn't necessarily mean he was in charge. Yet by the time they reached the waterfall, something between them was starting to click.

"So what do you think?" I asked Zeke. "Had enough, or should we keep going?"

"Oh, yeah. Let's go," Zeke said. "I'm really getting the hang of it."

"You don't know Sherman," I warned. "The best part of your day may already be over."

I prepared to repeat my Attack of the Growling Seagull performance, but as soon as I walked behind Flower and raised my arms, she automatically broke into a trot, almost as if the whole time we'd been talking she'd been waiting for me to shut up already and get back to running. Matilda and Sherman lurched off behind her, and the six of us cruised down the gravel road. We made it another quarter mile to the wooden bridge before Flower (of course) had to hit the brakes and thoroughly sniff it over in case some dramatic structural weakness had suddenly appeared since the last time we'd crossed it. Mika and Zeke didn't wait; they led their donkeys around Flower and quickly clattered across, inspiring Flower to follow.

Once Flower was safely on the other side, she towed me past the other two donkeys and surged off into the lead again. "See ya, suckas," I yelled as Flower and I vanished down and around a hilly curve. Maybe we'd underestimated Flower all along; rather than a twitchy baby who needed someone in the saddle to control her, she might be the best natural runner in the crew. I kept waiting for the trapdoor to drop and discover that, nah, it was all a fluke and we were now stranded a mile from home with a 400-pound block of cement, but except for her usual phobias, Flower seemed to be loving the opportunity to cut loose and leg it. The faster we ran, the harder Matilda and Sherman pushed to keep up with us, which could mean . . .

Which could mean . . .

I could *feel* the idea before I understood it, as if I'd spun the dial on a safe and heard tumblers clicking into place with no idea what was locked inside. Somehow I'd cracked the combination without even trying, and it took me a few more minutes of running down the road with Flower before I finally processed what we'd done:

We'd hacked Sherman's brain.

For months, we'd been struggling to train Sherman as my running partner, but maybe there was a simpler fix. Instead of changing Sherman, we could change the world around him. If Flower was a natural runner with fear issues, and Matilda was a fearless runner with abandonment issues, then maybe all we had to do was combine their strengths and weaknesses into one big Swiss Army knife of a support system. Whenever Flower was afraid, Matilda could step up. When Matilda lagged behind, Flower could set the pace. We had three weird donkeys on our hands, but together, they could give one another—and especially Sherman—all the help each of them needed.

I couldn't wait to put my new Donkey Multitool Technique to the test, so Flower and I stopped at the end of the gravel road for the rest of the gang to catch up. Flower was running out of her *mind*, and Sherman was doing fine with Zeke (according to Zeke, at least). So why not push on and see how far we could go if we switched the formation whenever one of the donkeys got finicky? As soon as I talked myself into it, though, Tanya's voice in the back

The beginning of a beautiful bromance: Zeke meets Sherman.

of my mind talked me right back out again. "*That's* your version of a good idea?" I could hear her saying, her left eye puckering with the pain of pointing out the obvious. "What happened to 'End on an up note'? Forget that?"

We also had a bigger problem: I still didn't know what kind of trouble Zeke was in. There was no way that straight-A, straight-arrow Zeke would suddenly abandon college unless he was in a serious jam. Had he gotten sick? Arrested? Was that mental-health issue from high school flaring up again? Any one of these possibilities was enough to make donkey running a terrifically bad idea. If Zeke was ill, stressed, or under house arrest, the worst place for him was somewhere deep in the Southern End, wandering for miles over lonely backroads with a feisty donkey and no cell-phone reception. Before I let myself get too excited, I needed to get Andrea by herself and find out what was going on.

"Okay," I told Mika and Zeke when they caught up. "Let's spin it here."

The donkeys frisked around, greeting one another with snorts and playful nips, but they straightened right out when they realized we were going home. Flower was so happy to be heading back to her pasture, she caused only a little kerfuffle when we reached the wooden bridge and snapped immediately back into a fast trot after I led her across. I thought about slowing her down when I saw Andrea walking toward us, but we were having such an inspired first run, I didn't want to leave Flower with any discouraging associations.

"You good?" I called to Andrea.

"I'm good!" she said. "I'm *very* good." She was beaming, and gave me a big thumbs-up. Only later, when I finally heard the whole story, did I find out why: for the first time in a frighteningly long while, Andrea had a feeling that her son's life might be out of danger.

17

It

A few weeks earlier, Andrea had been dead asleep when her phone rang at around eleven o'clock at night. She could barely understand the voice on the other end of the line. It was a young woman, very upset, saying something about Zeke. He was in the campus medical center at Penn State because he'd—what was that, cut himself? Yes, but it was more than a cut . . .

Andrea was wide awake now. She realized she was talking to Susan, a Chinese student who'd become friends with Zeke the previous year in their freshman chemistry class. Zeke had slashed his arm, Susan said, and then tried to hang himself from a rod over his doorway. Luckily, the rod broke loose from the wall in time, tumbling Zeke to the floor while he was unconscious but still breathing. He didn't know how long he was passed out, but when he finally came to, he remembered how horrible it felt to slowly choke to death. I'm no good, he thought. I'm such a loser I can't even kill myself. He called Susan for help.

Andrea lunged out of bed. She called her daughter, Ashling, then a junior at Penn State, and her husband, Andy, who was a five-hour drive away at work in upstate New York. Susan stayed in Zeke's apartment with him until Andy got there and took Zeke to the emergency room. Andy expected Zeke to be checked into Mount Nittany Hospital, which had a lot of experience helping Penn State students, but the mental health ward was full and the only available bed was in a secure facility more than three hours away. Because Zeke had threatened his own life, he was now in the custody of the state and had to be transported by the local constable. Zeke arrived for processing at five in the morning, exhausted and alone. His clothes were taken away and he was given hospital scrubs and assigned to a bed. After he had slept for only two hours, a hand banging on the door woke him up: time for group therapy.

Zeke stumbled out and groggily took his seat, wondering what had happened to him. The day before, he'd been a star science student at Penn State, majoring in physics and biomedical engineering with a specific interest in the mathematical modeling of brain circuits. This morning, he was locked in an institution somewhere in Reading, Pennsylvania, slumped in a plastic chair and surrounded by strangers who, one by one, were telling the saddest stories he'd ever heard in his life. The longer group therapy went on, the worse he felt. Zeke's fellow patients had been crushed by true horrors, tortured by abuse and addiction and imprisonment. And Zeke? He was well off and extraordinarily bright, a superb athlete from a fun-loving family who would do anything for him. What did he have to feel sad about? Instead of easing his depression, the session made Zeke feel spoiled and useless.

Afterward, he trudged back to his room to get some rest but found his assigned roommate was an older man with schizophrenia and sleep apnea who snored and "chewed tobacco like a beast," as Zeke put it, filling cup after cup with his mentholated tobacco spit. The room was so noisy and smelly that Zeke couldn't sleep, leaving him sluggish and withdrawn during the daytime activities. Zeke was desperate to get out, but his doctors, who didn't know about Zeke's involuntary all-nighters with the dip-chawin' snorer, saw only a suicidal young man who was still ominously quiet. They refused to sign his release.

Zeke's parents weren't giving up, and kept pushing their version of a plea bargain: Andrea was a licensed nurse, so if Zeke were allowed to come home, she promised she would treat him like any other patient and personally guarantee that he'd take his medication, attend daily outpatient treatments, and meet with a recommended therapist. After three days, the doctors finally agreed.

But the Zeke who came home wasn't Zeke. What had happened to the rambunctious dynamo who devoured books and double burgers, who had watched Richard Feynman physics videos for fun in high school and lived all summer in the backyard pool? All his curiosity and playful goofiness were gone, leaving behind a moody loner who didn't want to leave his room. Andrea and Andy didn't know what to do. If they pushed Zeke back toward his old life, would he try to end it again? And as a medical professional, Andrea had to take a hard look at herself and wonder:

Was this all her fault?

From the time her three kids were young, Andrea had worried about the risk of brain damage from contact sports. She knew that heading a soccer ball could rock a child's skull with nearly as much force as a helmet-to-helmet tackle, so before Ashling, Zeke, and Kelly were even old enough for Pee Wees, she was steering them away from fields and into the pool. When Zeke was in third grade and his older sister, Ashling, was in fifth, they were already swimming competitively year-round.

For an eight- and a ten-year-old, it was a grueling schedule. They had practice after school every day, with double sessions twice a week: on Tuesday and Thursday, they'd get up at four thirty in the morning; swim from five to six thirty, scarf down breakfast in the car on the way to school; and be back in the water for more laps before dinner. At night, they tackled their homework, then collapsed into bed. Kelly was only six years old, so she was mostly along for the ride, bundled into the car to go back to sleep while her brother and sister were churning out miles of laps before the sun came up.

The only ones busier than Andrea and Andy's kids were Andrea and Andy; when Andrea wasn't shuttling the twenty miles back and forth to the YMCA twice a day, she was working a full shift as a school nurse and attending night classes at the University of Delaware for post-grad degrees in nursing, health promotion, and health coaching. Andy, meanwhile, was commuting more than an hour each way to his job as a packaging engineer while simultaneously earning a master's in packaging science and strapping on his tool belt every evening to build out an upstairs rec room over the garage. Whenever Penn State had a home football game, the whole gang drove three hours to Andrea's beloved alma mater to tailgate. That's the way the Cooks operate: family first, full calendar, fully committed.

And for Zeke, perpetual motion was just what he needed; the pool was the only thing keeping him out of hot water. "I don't think he'd be a juvenile delinquent, exactly, but if it wasn't for swimming, he'd have gotten into more trouble," Andrea would reflect. Zeke and Ashling were both extremely bright, but Ashling, at least, managed to make that a virtue. Zeke was the kid who pestered the teacher with questions about chapters that hadn't been assigned yet, and waved his arm in the air while she was explaining the math problems to announce that he'd finished already. In medical terms, Zeke was a classic *proctalgia fugax*: a major pain in the ass.

"In school, teachers would get annoyed by Zeke because he'd finish the work in half the time and then start entertaining himself," Andrea would recall. "The more experienced teachers realized they had to keep feeding him extra work. In first grade, his teacher once let him go off by himself and read all day. Zeke loved it."

By eighth grade, Zeke was establishing himself as a formidable distance swimmer, and Ashling wasn't far behind. They bumped up from the Y to a national-level team at the University of Delaware, which meant tougher practices and a one-hour drive each way from the Southern End. "That was pretty savage," Zeke would say. "It got to the point where we did two five hundreds, two one thousands, and two miles"—nearly four miles of swimming, in other words, for a kid who was still in middle school. The new schedule meant they were doing homework in the car and not getting home till ten at night, and the intensity of the workouts left them constantly achy and run-down.

One year later, they were done. After devoting their entire adolescence to thrashing out laps, Ashling and Zeke told their parents they'd had enough. They quit swimming, and suddenly their lives became astonishingly . . . normal. Ashling, now a junior, finally had time to hang with her friends and stay out late for Friday-night football games. Zeke could play his first team sport, and promptly joined his high school wrestling team. Zeke knew nothing about wrestling and spent most practices getting slammed around on the mats, but compared to blowing bubbles two hours a day with his head underwater, wrestling was pretty fun. Ashling and Zeke loved their new life on dry land. Their grades were dynamite, they were making new friends, Ashling was National Honor Society vice president—

And then things fell apart.

Ashling was never a talker, but a few months after she and Zeke stopped swimming, she became quieter than usual. Except for going to school, she seemed barely to leave the house anymore. When Andrea asked if anything was wrong, Ashling just grunted to be left alone. Andrea had been a rebellious teen herself, so she recognized a defiant phase when she saw one. We'll just have to wait it out, she told Andy—until, during her senior year, Ashling said she wanted to kill herself. Andrea was stunned. How could she, a trained nurse who dealt with depressed students all the time, miss months of warning signs at her own kitchen table? She threw herself into getting Ashling help, and found a psychologist who had a knack for coaxing her to open up.

Too much stress, they all agreed. Now that she didn't have swimming to blame for missing homework, Ashling was feeling the heat to follow in the footsteps of her mom, who graduated second in her high school class and was accepted at Penn State main campus. But unlike her mom, who loved to study, or Zeke, who could goof off and still coast to straight A's, Ashling wasn't a naturally gifted student. She really had to grind for her grades. It was obvious that academic pressure was the cause of Ashling's depression—until the same thing happened to Zeke.

One year after Ashling's episode, Zeke was in the middle of a stellar junior year. He had perfect grades in five AP courses, and was still hanging tough on a wrestling team that was becoming a regional powerhouse. But he just felt so . . . *tired.* "I don't really know what happened," he later recalled. "You're fine, and then you're not fine." This time, Andrea sprang into action. Zeke met with Ashling's therapist, who believed that wrestling was squeezing Zeke with a level of stress and self-doubt that he'd never felt before. Zeke began medication and quit wrestling, and before long, he was back to annoying teachers at full strength again.

Zeke graduated second in his high school class, just like his mom, but he made her even happier by following his sister to Penn State. Zeke and Ashling were out of the woods at last, Andrea believed. Now that they'd identified the problem and found a strategy to defuse it, college was going to be a blast. Penn State was already their second home, so the only trouble they might run into was having too much fun. Still, it was good to know that Ashling and Zeke had each other to rely on. Just in case.

Luckily, Zeke got to Ashling while she was still alive. In November of her junior year, just two months after her brother arrived on campus as a freshman, Ashling swallowed a deadly amount of her depression medication. Before passing out, she thought one last time about never waking up again. She called Zeke, who raced to her dorm and rushed her to the hospital. Ashling spent the next three days vomiting and lapsing in and out of delirium, muttering nonsensically about someone stealing her Nobel Prize and suffering a seizure so severe that she was jolted out of bed and smashed her face on the floor, requiring a CT scan to check for cranial damage. When the pharmaceuticals were finally out of her system, she was remanded to the mental health ward for a week of intensive therapy.

Surviving an ordeal like that would leave anyone in need of a long rest. Not Ashling. She was released from the hospital in time for Thanksgiving at home, then went right back to school and knocked out her exams. "She was determined," Andrea would say. "I'm amazed at the way she went barreling through her junior year. She did really well." One of the doctors on campus was particularly devoted to Ashling's care, and that gave Andrea and Andy some peace of mind. As an added little safeguard, they got Ashling a cat named Finnegan and arranged for an emotional support permit so she could keep it with her on campus as a 24/7 stress-relief companion.

But most of all, Andrea wanted her kids to relax and not push themselves so hard. They didn't have to be such focused, disciplined, athletic overachievers—in other words, her. "Stress seems to make things tough for her," Andrea concluded, so from that moment on, Andrea was going to dial down the pressure and let her kids find their own way. Even when Zeke didn't bother to put his name in the housing lottery and missed out on a dorm for sophomore year, Andrea just took a deep breath and let it go. That was okay. They would find him an apartment. He'd like having a place to himself. No need to get stressed about it.

Zeke lay there, watching the clock tick over to 9:00 a.m. . . . 10:00 a.m. . . .

He was sick of lying in bed. He couldn't get comfortable, and he knew his parents were waiting for him at the tailgate. Just a few more minutes and he'd get up.

11:00 a.m. . . . Noon. . . .

Andrea's fingers were itching; the temptation to text Zeke was unbearable, but she didn't want to be That Mom. She'd already learned to back off after that time she'd visited Zeke's apartment and found he'd scrawled math equations all over the floor-to-ceiling mirrors. "Oh my god!" Andrea freaked. "He's turning into *A Beautiful Mind.*"

"C'mon, Mom," Zeke reassured her; it was just easier to study when he could see the formulas instead of having to look them up. After that, Andrea held herself back. Every single evening, she'd text all three kids to say good night. Ashling and Kelly always replied. Zeke never did. He's a young man, I get it, she told herself. I'll give him his space. She'd already seen that if her kids needed help, they would come to her. So she turned her attention back to her friends at the tailgate, and when they asked if Zeke was coming, Andrea shrugged and told them how crazy in love he was—with physics.

Outside Zeke's apartment, the rest of campus was raging. No matter how vicious the weather, no matter how early in the morning, when Penn State has a home football game you can count on two things: everyone is drinking, and they've already started. For the past decade or so, Penn State has consistently ranked as one of the hardest partying schools in the nation, even taking the No. 1 spot in 2009, and you only have to wander through on a Saturday night to understand why. Even Ira Glass, the gentle ironist who's heard every imaginable sin and psychosis during his twenty years hosting *This American Life*, was dumbstruck by what he found in Happy Valley:

IRA GLASS: So what's the wildest thing you've seen at a party at this school?

MALE STUDENT: Wildest thing? At a party? Somebody stripping completely naked and pretending to throw monkey feces as they were doing it. And—that was me.

IRA GLASS: That was you?

MALE STUDENT: That was me.

PENN STATE NEIGHBOR: One of the things you learn here in your yard is if you see a tampon, you have to get a stick and find the condom.

IRA GLASS: That is so gross.

Naturally, Penn State would rather be known for its fine legacy of scholarship than its new fame as a four-year baccalaureate booze cruise, but true to its academic mission, the school has been scientifically tracking just how much shot-slamming is really going on. What the administration itself found is that each Friday and Saturday, *three out of every four* undergraduates are busy pounding drinks. So if you're one of the lone Nittany Lions who's not such a fan of football, Fireball, and feces-flinging, every weekend can make you feel like a castaway who washed ashore on an island overrun by rampaging marauders.

Because of Zeke's goof with the student housing lottery, he spent his sophomore year in a studio apartment next to "Beaver Canyon," the notorious downtown scream tunnel on East Beaver Avenue where students throng by the thousands after football games (if you saw footage of the Penn State riot after coach Joe Paterno was fired in the wake of the Jerry Sandusky sex-abuse scandal, Beaver Canyon is where students tipped over a TV news van and tore down lampposts). All Zeke had to do to join the fun was step outside his apartment, and he'd immediately be swept up in a sea of micro-dresses and beer bongs. His apartment pulsed with the shrieks and music echoing below his window—while he lay there, wondering why he felt too tired to get up but too tense to sleep.

Zeke couldn't figure out what was wrong. Freshman year had been *amazing*. He'd hit it off with his roommate, an astrophysics major who was happy to kick back with him in their room and geek out about science, and he did so well in his classes that he soon advanced from coursework to hands-on lab research. Penn State's biomedical research facilities are so well known that even the superhero universe held them in high esteem; Zeke could point out that of all the colleges in the world, Dr. Bruce Banner* chose

* Before going on to battle intergalactic evil as a member of the Avengers, Dr. Bruce Banner (aka the Incredible Hulk), was an undergraduate at Penn State in the original Marvel comic series.

Penn State for his pre-Hulk education. Zeke was especially fascinated by the motor proteins, the neurological superstructures that convert cellular energy into mechanical force, and soon he was hanging out with scientists whose work might someday stop Alzheimer's and cancer in their tracks.

"My self-esteem was at an all-time high," he'd say. "First year at Penn State, I was knocking it out of the park." Zeke felt so sharp, so clicked-in and on his game, that he decided the only thing slowing him down was those damn antidepressants. How can you hook up with girls when your body is puffy, or impress professors when your concentration drifts? His solution was to quit taking the meds. Before long, his six-pack was popping again and his brain was a laser. Depression was just a phase, he decided, and that phase was behind him.

"Then—I don't know what happened," he would recall. "It came bubbling up."

It felt like a backpack full of stones, a weight he could barely lift and couldn't shuck off. *It* came out of nowhere, paralyzing him as he lay in bed. And the only thing that would make it go away for good, Zeke decided, was that metal rod over the bathroom door. . . .

After Zeke was rushed to the psych ward, Andrea was so preoccupied with the fight to bring him home that she could barely think of anything else. "All the professionals said he needed to be in the hospital," she would later say. "But when we saw him, we realized it was making him more depressed." Only after she had Zeke back in his own bed could she find the mental energy to finally confront the question that had been nagging at the back of her mind: Was she to blame?

It wasn't guilt she wrestled with. It was science. "I look back and have to wonder," Andrea says: Why did depression attack two of her children so aggressively, while the third was unaffected? Kelly, her youngest, was only two years younger than Zeke. She grew up in the same household, went to the same high school and college, hung around with similar friends. Kelly and Ashling even had identical tattoos, an outline of the family vacation spot at Lake George. But at one crossroads, their lives veered apart: during those mornings when Zeke and Ashling were in the pool, Kelly mostly slept in. "Ashling and Zeke were both swimmers, so they were used to these endorphins. This *high*," Andrea recalls. "And then suddenly, it stops."

So what, now swimming is bad for you? How was that possible? Exercise, as Andrea had been told a million times in nursing school, is the Great Healer, an all-purpose wonder drug that can improve everything from digestion to depression. Exercise was medicine before we *had* medicine, and because it's so vital for keeping our internal machinery humming, our brains evolved to give us an attaboy! whenever we break a sweat by rewarding us with, basically, a squirt of Grade A pharmaceuticals. When you work out, the pleasure centers in your brain are flooded with endorphins and dopamine, the "happy hormones" that make you feel the way Dwayne Johnson always looks: relaxed, strong, confident, intelligent. Brain opioids are so powerful that if you exercise, you can lower your "mental health burden" by nearly 25 percent and enjoy a much higher ratio of positive mental health days—a whopping 43 percent—than non-exercisers experience. Let that sink in for a second: just by goofing around a little on your bike, you can nearly *double* your happiness. For free. If you put results like that in a pill, it would outsell ice cream.

No wonder pretty much everyone who walks out of SoulCycle seems like they summited the Himalayas and got chest-bumped by a bodhisattva. They're not just relieved that the hour is over; they're floating on a cloud of their own naturally produced party drugs. Brain opioids are so potent that they even make lab rats eager to work: after rats become used to exercise, they'll perform a monotonous job, like pushing a lever over and over, for a chance to get back on the running wheel and be rewarded with a fresh surge of dopamine.

But any drug with that much firepower, Andrea knew, even a drug produced by your own body, has just as much capacity to lay you out. As a school nurse, she eyeballed hundreds of kids every day for hints of drug use. She was also a dedicated athlete herself, so when she combined her knowledge of narcotics and her own experience with exercise, she had to wonder: If you spend half your life getting a daily superdose of dopamine, what happens when you suddenly quit? Do you go through withdrawal, like anyone else kicking a chemical habit? Judging by Ashling's and Zeke's symptoms—insomnia, anxiety, mood and weight swings, acute depression—it certainly seemed a lot like they'd gone cold turkey. That hypothesis could explain why competitive athletes, who work out more than anybody else, may be nearly twice as vulnerable to depression as non-athletes. Did they pass some dangerous, invisible cutoff at which the cure became an affliction?

That's exactly the question that researchers at the University of Bonn in Germany set out to answer in 2008. Until then, the only way to peer into the workings of brain hormone levels was to subject volunteers to the torture of a spinal tap. But thanks to neuroimaging breakthroughs, the Bonn team was able to sidestep the giant needles and, instead, inject ten volunteers with a tiny radioactive tracer that could be tracked by a scanner. Previous experiments had shown only minor dopamine reaction after thirty minutes of running, so this time, the volunteers were asked to hammer the treadmill for a solid two hours. When they finished, their brain opioids were calculated. Not only had all of the runners' opioid levels increased significantly, but the Bonn researchers made another surprising discovery: the better each runner felt, the more dopamine was found in the spinal fluid. Rather than an on/off switch, the hormone was acting like an intoxicant: the higher the volume, the happier you feel.

And the farther you fall. No wonder athletes have a hard time adjusting to life off the court and out of the pool. When researchers dig into the mental health histories of athletes, that's exactly what they discover: the most perilous moment is when athletes face "injuries, career termination, decline in performance, or a catastrophic performance." There's always been a dim awareness of a post-glory-days slump, but it was written off as nothing more than an ego bruise, an overdue dose of humble pie for sports heroes now facing life as mere mortals. Instead, it could be something far deadlier: a dangerous chemical imbalance caused by a sudden drop in dopamine. And the most at-risk population? "Solo athletes," reports Professor Jürgen Beckmann, chair of sports psychology at Germany's Technical University of Munich and lead author of a depression comparison study. "We have very high prevalence rates in swimming, for example."

Michael Phelps isn't surprised. "The world knows me as a twenty-eight-time medalist. But for me, sometimes my greatest accomplishment was getting out of bed," Phelps has said. "When depression hits, it can become debilitating and feel like nothing really matters." Sometimes he just lay in the dark, longing to die. "For me, getting to an all-time low where I didn't want to be alive anymore, that's scary as hell," Phelps said. "I remember sitting in my room for four or five days not wanting to be alive, not talking to anybody."

Phelps never knew that one of his closest friends, eight-time Olympic medalist Allison Schmitt, was struggling with the same despair. Like Phelps, she suffered in secret until tragedy struck: in 2015, Schmitt's cousin, high school basketball standout April Bocian, killed herself one week after her seventeenth birthday. Schmitt was tormented by the thought that if she had spoken up and gone public with her own depression, her cousin might have reached out to her for help. "She was in such a dark place and so isolated and felt so alone inside," Schmitt has said. Since then, Phelps and Schmitt have become advocates for mental health awareness and such a tight mutual support team that Schmitt now lives with Phelps and his wife and son in their home in Arizona.

One college coach was so heartsick over April's death that she launched a foundation in her honor even though April never played for her. Suzy Merchant was recruiting April for Michigan State when she heard about the tragedy. Now, every year on April's birthday, Coach Merchant gathers hundreds of young women on Michigan State's campus for empowHER, a weekend retreat aimed at preventing the loss of another struggling young woman. April's mother leads them in singing "Happy Birthday" to her daughter and reminds them they have to look out for one another. "Just because you're feeling better one day doesn't mean it's gone, doesn't mean you're healed," Allison Schmitt reminds young athletes. "It's something you have, and have to live with for the rest of your life."

Andrea felt her heart sink. The more she learned about the possibility of the dopamine-depression connection, the more she was wracked by guilt. She was the Tiger Mom who pushed her kids to swim. She was the one who insisted they get their butts in the car when they wanted to skip a day. And now look at what the Bonn researchers were reporting: hormone levels were being jacked up by a two-hour workout. Two hours—the same as Zeke's and Ashling's practices. For roughly half their lives, they'd begun and ended every day with a blast of a powerful mood enhancer.

Okay, Andrea decided. *Enough.* Beating herself up wasn't going to help her kids. She had the rest of her life to worry about blame; right now, she had to focus on remedies. One thing that had helped Ashling was getting her a cat, and now Andrea understood why: pets are a great way to spur the release of oxytocin, a hormone that functions much like dopamine. The Cooks found the perfect oddball cat for Zeke—a one-eyed creature he instantly named Schrödinger, after the father of quantum mechanics—but Andrea knew that with Zeke's restless energy, he needed more than a plaything.

In her gut, Andrea knew her son should be out there exercising again, feeling the sun on his bare back and coming home happily exhausted, but the prospect filled her with dread. What if he pushed himself back into dopamine dependence? And did she really want him disappearing into the woods for hours at a time on his own? She wasn't even comfortable letting him go to therapy by himself. "I don't know if he knows, but I followed him down to his first outpatient treatments," Andrea told me. "I went to make sure he was going, making the right choices. You always worry because he's smart enough to know what to say. Anyone with depression has the ability to hide their underlying feelings."

One day, Andrea and Zeke were taking a walk together when he asked if it would be okay if he called her friend Chris. Did she think I'd mind? Because as long as he was going to be home for a while, maybe I could take him on a run and talk about how diet and fitness affect depression. Andrea was so excited, she had her phone out of her pocket and was texting Mika before Zeke even finished his thought.

"For the first time, I felt he was on the right path," Andrea says. "He was finally getting outside himself." She was a little jolted the next day when they came to the house and suddenly learned that the kind of running we had in mind involved an animal with hooves, sharp teeth, and a seriously troubled past. But when Andrea saw her boy loping alongside this damaged creature, the weight on her chest finally began to ease. "Once he started hanging out with you guys, it filled my heart with joy," she later told Mika and me. "Sherman became his purpose. He'd found someone else who needed healing."

18

Plan C

Sports don't build character. They reveal it.

—CURTIS IMRIE, burro master and cowboy philosopher

Shit," I muttered. "He's here."

"Already?" Mika joined me at the kitchen window, where I was watching a red MINI Cooper rumble up the driveway. It was miserable outside, a dank March morning with a hint of something wetter and worse on the way. Mika and I had just decided that all of us, donkeys included, would be a lot happier if we bagged our run and stayed inside. It was only eight thirty, which I thought was plenty of time to wave off Zeke before he was due to arrive at nine, but as I got up for the phone, he was pulling in.

"Wow," Mika said. "Even if he was just getting out of bed now, he'd be early."

I rapped on the window and waved for Zeke to c'mon inside, then slid a frying pan onto a burner. Zeke loved a second breakfast, as we'd learned over the past few weeks, and his cast-iron belly let him run immediately after packing one away. While he was on the porch saying good morning to the cats, I cracked two eggs into the pan, sliced a tomato, put a fistful of bacon under the broiler, and reheated a triple espresso with a thick dollop of fresh cream from our neighbor's Jersey cow. I reached for the rye bread to toast him a slice, then remembered and put it back. Zeke was cutting out all processed carbs and sugars to see if lowering his blood sugar would help stabilize his depression. Ketogenic diets had proven effective for epileptic seizures, and while as yet there wasn't much hard science behind their usefulness for depression, Zeke was making himself his own lab rat. For a young guy with a roaring appetite who lived in the heart of Amish sticky bun country, he'd been amazingly disciplined. And even though we didn't talk about it much, I knew why.

How many times have we been shocked when someone strong, successful, and adored is suddenly snatched away by depression? David Foster Wallace, Anthony Bourdain, Kate Spade, Robin Williams . . . they all had family who loved them and access to the finest resources in the world, yet the disease still overwhelmed them. April Bocian, the young basketball star, was lost even though her extraordinarily attentive parents had already sought psychiatric help for her by eighth grade. "She would start off with school doing very well and excited about basketball, and by October, November, she would be exhausted and it would be very hard to get her up for school," April's mother, Amy, has said. "Then it would just snowball from there." No matter what you try, Amy learned, no matter how excellent the care, depression is a black wave that can surge when you least expect it and sweep your loved ones away.

Zeke got the message. I don't know how he learned it, whether from his own generosity of spirit or the shock of waking up on the floor with a noose around his neck, but he was smart enough to grasp that he didn't have all the answers, and neither did his doctors. Zeke knew he was trapped in a dangerous maze, one that had almost killed him already, and if he was going to find a way out, he couldn't stop searching. "I'm trying everything," he'd told me. "This is the second time I've been thrown from this horse."

"So," he asked, after the scent of fresh eggs snapping in butter had lured him away from the cats and into the kitchen. "What's up for today?"

"Until you showed up we were thinking about a day off, Sarge," I said. "Now that we're going to get wet anyway, we might as well try some creek crossings."

It was a little early in the game to be worrying about water hazards; I knew that. We'd talked with Zeke only about *maybe* going to Colorado, and *maybe* he'd like to come, and that's as far as we'd gotten. Actually, we were backsliding; my half-assed plan from a few months ago had crumbled so much, it was now barely a quarter-ass. One by one, the things I'd thought I could count on were falling apart. Could we still use Tanya's stock trailer, and would she drive? Was there any way Sherman would race on his own, or did we have to haul two donkeys? *Or three?* Would Mika seriously step up for thirty miles of high-mountain hill running? Would Matilda? I had no idea. But one thing I knew for sure: somewhere on that racecourse, cold-water creeks were waiting. If we ever got to the World Championship, we had to be ready.

"Okay. Be done in a sec," Zeke said, standing up with half the food still on his plate, shoveling forkfuls into his mouth as he headed toward the sink.

"Relax," I said. "Sit and chew." So far, my initial concerns about Zeke had proven completely misguided. No matter what I suggested for that day's workout, no matter how cold or muggy it was outside or how ball-breaking the donkeys were that day, Zeke never kvetched or second-guessed. Maybe it was his swim team training kicking in, but he seemed to have looked around at the three of us and decided if Mika and I were the coaches, he must be team captain. When I made a point of complimenting him for never showing up late or missing a workout (even when we wanted him to), he replied, "Yes, I realized you were relying on me to supply the punctuality and consistency." He wasn't being snotty; he was being Zeke.

. . .

Sherman came right to the gate, not put off a bit by the nasty weather, while Flower and Matilda lingered in the shed before reluctantly schlumping along behind him. Flower fussed while we haltered her up, but only when we started jogging up the road did she really make us pay for coaxing her out into the rain. We were running in a tight pack, with Zeke sandwiched between Flower and Sherman, when Flower unclenched her butt for a few bugle calls. Her farts were long, loud, and rhythmic, each one a human-rights violation timed to the tempo of her stride. Poor Zeke was trapped in the blast zone, caught between Sherman on his left and an electric roadside fence on his right.

"Hoo, *mama*," he winced, waving his hand in front of his nose. "Crop duster."

Zeke didn't escape the line of fire until we reached the turn-off onto the gravel road. As soon as he could wiggle free, Zeke dropped his rope and sprinted in front of Flower. He raised a hand in warning—*incoming!*—and uncorked a face-melter of his own. When it came to guerrilla ops, that second breakfast really paid off.

Zeke flashed Flower a peace sign. "Truce?" he offered.

"Oh my god," Mika groaned. "We need more women in this group."

Personally, I was elated. I'd been scared to death about making a mistake with Zeke, like pushing him too hard or missing a sign that he was in trouble, but if he was playing Call of Duty in a hit-and-run fart war with a six-foot donkey, I had to believe he was on solid ground. Not to mention, hilarious.

"My man!" I applauded. "Now holster that weapon."

Zeke scooted back to Sherman and picked up the rope, and the six of us settled into our pace again. At the end of the gravel road, a break in the brush led to the creek, and that's where I called the gang to a stop. I led Flower through the trees and down to the edge of the water. We didn't have an emotional-support goat with us, the way we did that time we got Sherman to wade in behind Chili Dog, but I was banking on getting the same results by using the same body language: I'd stride straight into the creek, never looking back, using confidence and purpose to transmit my Alpha Dog Authority up the rope to Flower. If she kept moving, the other two donks should fall into formation behind her.

The creek was only shin-deep and flowing smooth, no more than six strides across. I walked in, calm and easy. Just to be safe, I gripped down tight on—

Air. Nothing but air. The rope was gone, whipping out of my hands as Flower backtracked up the bank, scuttling behind Sherman and Matilda for protection until Mika stomped on the dangling rope and brought her to a stop.

"How about I try?" Zeke asked. Sherman was still next to the stream; he'd never budged when Flower zoomed off, watching her go as if they'd never met before. *Hmmm* . . . Did that mean our Chili Dog water-skills course had really sunk in and the new, all-terrain Sherminator now thought Flower was acting like a wuss? Or was he just pooped from the run and taking a little Sherm-time? I weighed our options: Zeke had no experience in taking donkeys into water, but at the moment, I had no donkey.

"Yeah, give it a go," I agreed. "Plow on through and see if he follows."

Zeke was perfect. Somehow, he sensed exactly how much force to put on the rope and where to position himself alongside Sherman's head. I couldn't have done much better myself, and told him so—right after he got yanked off his feet. Sherman butt-squatted and jerked his head, tugging Zeke into a desperate, staggering dance and into the drink. Zeke scrambled up and tried again, this time straining on the rope like a rock-snagged sailor and resorting to the one word that never works with anyone: "C'mon, Sherm," Zeke implored. "C'mon, c'mon. Come *on*!"

"Mika," I called. "You're up." To Zeke, I said, "Don't turn it into a fight. You'll lose, and you'll only teach him how to beat you. On to Plan C."

Mika handed me Flower's rope and led Matilda to the bank. "Here we go, baby," she crooned, wading across. Matilda hesitated, then splashed in behind her. Sherman was still busy defying

Zeke and hadn't uncoiled from his on-this-hill-I-die crouch, so I brought Flower on deck. I was still double-wrapping the rope around my fist when I heard Zeke say, "No, no, no. Oh, no. Watch out!"

I whirled just as Flower launched, blasting off the bank in an attempt to clear the stream with one giant leap. Everyone could tell right away that it wouldn't work—everyone, that is, except Flower, who only had eyes on Matilda, and Mika, who wasn't looking.

"MOVE!" I yelled.

Too late. Six hundred pounds of flying donkey came crashing down in midstream, missing Mika by inches. Mika, stumbling, managed to find her footing and recover. Not Matilda. She ripped out of the creek, galloping for her life, probably assuming only a she-bear on the loose could make Flower freak out like that. Hot on her heels came Sherman, who didn't know what was going on but was afraid all the commotion meant he was about to be abandoned. Mika and Zeke instantly released their ropes and got out of the way, not keen on being dragged across the rocks. When the three donkeys made it to dry land, they stopped and looked back, as if the three soggy humans were the ones holding up the operation.

"I hate to say this," I said, as we slogged across. "But we should take them back across right away, so the lesson sinks in."

"Hold on to Flower this time," Mika replied.

"It's not that I'm *not*—"

"Yeah, okay. Just hold on to her."

"Definitely."

And I tried, but Flower cannonballed again anyway. This time Mika was ready and sidestepped out of the way before Flower was even in the air. On our third trip, Flower pawed the bank nervously a few times before wading in cautiously. By our fifth crossing, Flower and Sherman were both traipsing behind Matilda with barely a tug.

"We got this," Mika said.

"At least until tomorrow," I agreed, ready to call it a day. We'd been out for nearly two hours, and even though we hadn't run very far, we'd had a huge day. When we set out, I wasn't sure how Zeke would handle his first stressful challenge. I remembered a parkour coach who once told me that every group of guys he taught had three types: the joker, the show-off, and the explainer. Zeke was none of those. He stepped up when needed and back when not. He didn't get upset when he failed and didn't lay blame, not on himself or us or even Sherman. We were tired and soaked to the skin, but all of us—humans and donkeys alike—had tested each other and discovered we could be trusted.

"Ready for lunch, Zeke?"

"Starving."

We clambered out of the creek, shoes squishing, and gave the donkeys plenty of slack so they could race one another up the gravel road toward home. As soon as they were through the gate, they headed for the barn, scattering the sheep and goats, who were gazing out at the rain, thrilled, I'm sure, that they weren't tapped anymore for donkey training. While Zeke and Mika stowed the ropes and halters, I went inside the house and got bacon started for BELTTS (bacon, egg, lettuce, tomato, and Tabasco sandwiches).

Mika and Zeke came in while I was dialing the phone. "I'm going to find out a little more about those Colorado creeks," I lied. Really, I just wanted to brag. In the years since my first and only burro race, I'd become friendly with Hal Walter, the legendary champion who was still winning races in his fifties. In his day life, Hal was a reporter and freelance editor who wrote about his adventures on Colorado's Front Range, and he'd been extremely encouraging when I first reached out to tell him what I was attempting with Sherman. When I originally ran into him in Leadville, I thought he was cold and kind of a dick, but I'd discovered that was just Pre-Race Hal, who's still so serious about the sport that thirty-five years of winning hasn't calmed his nerves. Post-Race Hal, on the other hand, is a teddy bear.

"You getting your miles up yet?" Hal asked when I got him on the phone.

"Little bit. Today we were focusing on stream crossings."

"Good," Hal said, approvingly. "Good for you. All that roaring water can just kill your race."

"Roaring water?" I repeated.

"*Roaring water?*" Mika and Zeke both blurted.

"How much, um—how big are these creeks?" I asked. Mika and Zeke were now both staring at me intently.

"Depends on snowmelt and weather," Hal said. "You can get 'em waist-deep, and I'd say, oh, thirty, forty foot across."

"Thir—" I began, before feeling Mika and Zeke's eyes on me. "Okay! Looks like we've got a little more work to do." I thanked Hal and hung up, then hurried over to pull the bacon and toasting bread out of the oven. "Whew, just in time," I said, but Mika wasn't falling for that dodge.

"What's that about 'roaring water'?" she asked.

I downplayed the phone call, busying myself with the BELTTS and agreeing that, yeah, the race was a little wilder than I thought, but why worry? If we decided to go ahead with this, we still had plenty of time to prepare. Privately, I was worried. I was leading my wife, a struggling young friend, and three vulnerable animals into the mountains with no way to get them back out if we ran into trouble. There was a lot I had to find out, and it was time I got started.

19

You Rarely Win, But Sometimes You Do

Done right, this sport leads you to less of that mañana attitude. It's here. Now. Real. It's the secret of the pioneers: persistence, patience, and prairie.

—THE EVER-QUOTABLE CURTIS IMRIE

WHAM!

The slam of a screen door made my head jerk up. I'd just driven down a lonesome road near Colorado's Sangre de Cristo Mountains and pulled into Hal Walter's dirt driveway. When I got out of the car, some kid I'd never seen before banged out of the house and came barreling at me like he was dead-set on taking me out.

"Son," I heard someone behind him shouting. "Son!"

Nothing about Sonny suggested he'd be any easier to stop than a ballistic missile. He was a kid on a mission, his fists clenched and legs pumping, but when he was a few steps away I realized by his angle of attack that his guidance system wasn't necessarily trained on me; it was aiming more like left of center. *Oh, shit.* Even worse. Now I understood why the voice in the house was so panicky. Behind me, three mammoth creatures were pressed up against a paddock gate. Colorado donkeys aren't a bunch of Flowers and Shermans; they're a whole different breed, descended from wild stock that survived by stomping mountain lions to death. Hal watched a buddy get his teeth kicked out, and recently Hal himself got a nasty chomp. If a master burroman like Hal wasn't safe in that corral, then clearly this kid's mad dash was heading for trouble.

"Hey, hold up," I called, but Sonny darted past and leaped for the gate. I braced for carnage—

And found Hal reaching out for a handshake.

"Good trip?" he asked. "Harrison, come over and say hello."

So that was Harrison—aka "the Blur." I knew a little about the Blur from my calls with Hal; I knew he was eleven years old and "neurodiverse," a term Hal prefers to "autistic" because it "opens the possibilities," as Hal puts it, "and sets aside stereotypes." When Harrison was still little, his parents discovered he was a musical and mechanical whiz; he sang with perfect pitch, took up the piano with ease, aced computer games, and loved to take apart and reassemble locks and clocks for fun, the trickier the better. On his best days, Harrison will serenade his mom with Mumford and Sons songs, and on his worst . . . well, sometimes that's the same day. During a middle school cross-country meet, Harrison was running beautifully until the noise of cheering made him flee and hide in the bushes. He wouldn't come out until a kind woman with a golden retriever used the dog to coax Harrison to the finish line, which he crossed by rolling over and over like he was on fire. Harrison actually has terrific leg speed and the endurance of a horse—hence, "the Blur"—but it was impossible to predict from one meet to the next if he would run hard, or high-five everyone on the sidelines and finish last, or wade furiously into the crowd, swinging his fists, until Hal and his burro-racing compadres could grab him.

"Harrison, come on over," Hal called again. "Let's introduce Laredo."

Harrison Walter, aka the Blur

Amazingly, the mammoth donkeys hadn't even flinched when Harrison landed with a bang on the gate. "They understand him," Hal told me, "and always have." When Harrison was having a really tough afternoon as a youngster, Hal would scoot him outside and onto Laredo's back. Within seconds, boy and burro would be ambling along, Laredo's ears waving as Harrison sang "Yellow Submarine." Few people alive know as much about burros as Hal and his wife, Mary, and thanks to the life they've created out here on the front range with Harrison, no one can match their hands-on experience when it comes to pairing these big animals with the neurologically diverse. I wanted to soak up everything they could teach me, especially about whether it was smart, or stupidly risky, to expose a vulnerable kid like Zeke to such a frustrating and potentially demoralizing sport. Sherman is adorable, but I'd learned firsthand that he needs only a minute to skyrocket your blood pressure and bottom out your self-esteem.

"Yup, I hear you," Hal agreed, as I told him about my own Gang of Three while we stood at the gate, admiring his trio of racing burros. "Weird how much you love them when they can make you hate yourself."

Over the years, Hal has been kicked, bitten, rope-burned, abandoned, and flat-out bewildered by burros. Mary, an exceptional athlete herself who won three straight World Championships in the early 1990s, was badly battered when her leg got tangled and a spooked donkey dragged her down a rocky trail. But all of that has been forgiven. Slate wiped clean. Because when Hal and Mary were at their lowest—when they realized their beautiful little boy would struggle for the rest of his life—the donkeys came through in ways they never could have imagined.

"I'm not melodramatic," said Hal, kind of unnecessarily for a guy who looks handcrafted from hardwood and saddle leather. "But the best parts of me I owe to those animals."

Hal and Mary met when they were just kids themselves, teenage freshmen at the University of Colorado in Boulder. Between classes, they hiked and ran and skied together, and when they graduated and embarked on careers—journalism for Hal, nursing for Mary—they chose their first home because it had enough pasture for Jumpin' Jack Flash, the burro Hal had just adopted. Hal got his first taste of burro racing while still in college, when his phone rang out of the blue one evening with a call from Curtis Imrie. Curtis needed an extra set of legs to help with some burros he was prepping, and he'd been told Hal was a strong trailrunner. Hal was intrigued—and terrible. He finished his first race in last place, and spent the next eighteen years trying, and failing, to beat wily veterans like Curtis before finally breaking through to win an astonishing seven World Championships. Burros are probably the only reason he bothered to keep running. "Road races are so tedious and repetitive," Hal told me. "But with burros, you're totally focused. You never want to quit, because you feel like you're one step away from solving this amazing puzzle."

In 2015, Hal caught a stomach bug a week before the first race of the season, a little nine-miler in Georgetown, Colorado. He was too sick to run, but he'd promised he would drive a stock trailer with burros for two of his friends. Sleet was battering the windshield as he crossed the mountains, and that's when he realized he had a flat tire. He pulled onto the shoulder, coughing and sniffling, and herded the burros out of the trailer. With freezing fingers, he changed the tire, wrangled the burros back on board, and sped off over icy roads to make it to the race on time. He arrived, soaked and shivering, only to find that one of the guys who'd asked him to haul a burro all that way had changed his mind and wasn't going to run. To hell with it, Hal decided; I'll race myself. He dug a pair of running tights out of the back of his truck and jogged to the line still wearing his heavy Filson barn coat. He was going to shuck the coat at the start but was still so cold that he left it on. Not surprisingly, Hal hit the halfway turn in last place. Leading the field was Justin Mock, a thirty-two-year-old Denver speedster who'd come to burro racing along a weirdly appropriate path: he'd lived on a ranch as a boy with a nasty pet zebra that chased him around, and that backyard survive-and-evade training eventually helped him become the top American finisher in the 2010 London Marathon *and* the fastest human ever to run the Bolder Boulder 10K in a gorilla suit. In the Georgetown burro race, Justin was heading toward the finish in first place when he looked back and saw some crazy guy in a flapping canvas coat bearing down on him. Justin sprinted. When he glanced back again, the madman was gone . . . only to pop up in the lead. Hal had waited for Justin to look left, then scooted right to win. He was fifty-five years old.

"I wouldn't say burro racing *consumed* me," Hal protests. But, yeah, okay; he did quit journalism for a while to pursue a career as a professional burro racer, hauling his stock trailer across the Southwest as he and Curtis pool-sharked from town to town in pursuit of cash prizes. And fine, he'll admit it was the burros that finally got him to propose to Mary, inspiring him to take a knee next to the tub when she was up to her neck in bloody bathwater, cleaning her wounds after that brutal tumble on the trail. *Burros*

Hal Walter proving that an old burro racer can still beat elite marathoners

don't even stick with the sport as long as Hal; he's been in the game so long he's had to replace eight burros he felt had gotten too slow or too old. That's right; donkeys have the longest work life of all equines, and it's still shorter than Hal's.

But all of it—all those freezing mountaintop miles, the mountains of hay he's had to buy and stack, and the miserable hours he's spent tromping through storms in search of lost strays—it all made sense when Harrison was born. Hal thought he'd gotten pretty good at training donkeys. He never realized they were training him.

Mary was the first to suspect something was wrong. "She was the *only* one, to be honest," Hal said. Mary, listening to her husband tell the story, glanced over to make sure Harrison was still engrossed in his video game, then nodded in agreement. When Harrison was barely a year old, she began to notice things that worried her—and her alone. "Hal just didn't want to believe it," she said.

We've come inside the Walters' charming farmhouse, with its sweeping views of the grassy-green Colorado mesa stretching to the red-streaked mountains. Earlier that morning, Mary had been having a tough time with Harrison; now that he'd grown taller, stronger, and louder, his outbursts were harder to manage. At the moment, though, there was no hint that their lives could be anything other than perfect. Harrison is a terrific-looking kid, a lean young rascal with a tousle of blondish curls and a playful gleam in his eye. Just like Hal, Mary is sun-browned and fit, gracious and wry, a thoughtful listener who's quick to laugh and add a smart insight. You wouldn't suspect trouble was lurking in this family—and Hal didn't, for a very long time, even when Harrison almost died.

Twice when Harrison was little, Hal had to leap into action when he was choking, compressing his chest with baby Heimlich pushes until food erupted from Harrison's windpipe. *Just accidents*, Hal muttered when Mary insisted that eating problems are an early red flag. So were Harrison's other quirks, she pointed out. Like repeating the same word, over and over. And obsessively opening and closing doors. And the way he'd dart around wildly, not aware that he was heading straight for the furniture. And what about his extreme skin sensitivity? "He would not wear gloves of any type," Hal conceded. "A loose string in a sock could cost you an entire morning with screaming and tantrums."

Over Hal's objections, Mary had Harrison examined by specialists. During one session, Hal began to fume when the therapist kept referring to Harrison's "autistic-like" behavior. "My inner cowboy just wanted to toss the psychologist out the front door," Hal grumbled. But that night, he heard Mary crying privately in despair, and he finally woke up to what he had to do. It was time to stop fighting his son's situation, Hal realized, and start helping. But how? He was a wrangler, a grab-it-and-heave-it kind of guy. Why the heck else would he still be busting his hump as a knock-around reporter and ranch manager when he could have settled into an office job years ago? Hal bet that if those psychologists had gotten a look at him, they'd've been writing him a prescription too. How was he going to help Harrison when he could barely sit still himself?

Yup, Curtis Imrie agreed when Hal opened up to him. If this autism spectrum is real, you're probably on it.

But so am I, Curtis added. Seriously, think about all our burro buddies, he reminded Hal: Tom Sobal, who was such a loner he liked to race under fake names; and Clint Roberts, who got knifed by his girlfriend's ex; and their dear lost pal, Rob Pedretti . . . They were wild and wonderful, every one of them, the best of companions and marvelously compassionate burromen, but c'mon: no one would ever argue that they weren't a little off-kilter.

Maybe that's how we self-medicate, Curtis mused. We throw a halter on our partner and head for the high places, climbing toward the sky until something reboots our brains and makes us feel right again, whether it be the whispering pines, or our hammering hearts, or the steady breathing of a gentle companion.

We weirdos are the last custodians of lost wisdom, Curtis believed. We're like the square pegs of the past, all those mystics and miners who roamed the world with donkeys, because we know that when it comes to round holes, four hooves is the way we fit in.

After all, look what it did for young Ben Wann.

Three hours away, near Denver, ten-year-old Ben had suffered an epileptic seizure so severe that it took a week in the hospital before he could speak again. Ben's parents, Brad and Amber Wann, were terrified that the next attack could kill him. Nothing seemed to help, but anyone who's met the Wanns knows that if you really want to fire them up, tell them there's nothing they can do.

Brad is a big, bearded, burly-looking bruiser who rocks your world for a living by installing booming speakers inside your furniture. Amber is warm-hearted and sweet and half Brad's size, but when it comes to protecting their kids, it's a coin toss who you'd want in the fight. Take medicinal cannabis, for example: when

Amber found that hemp oil could be effective, she made sure Ben always had some on hand even after the school called children's protective services on her. Amber took her fight all the way to the state capital, and didn't back off until the legislature legalized cannabis medicines in the school system.

No wonder that soon after she heard about equine therapy, Curtis's phone was ringing. Curtis had never done anything like this before. He was an aging bachelor who lived, very quietly and happily, deep in the woods with his girlfriend and his half-wild herd along a washout that's more ruts than road. But what the hell; if the Wanns were up for a long drive and a little sweat, they were welcome to visit. Curtis had the Wann family come over for Thanksgiving, and in the middle of the meal, Ben suddenly spasmed into a seizure. "Ben has no memory of it," Amber says. "But Curtis—oh god, it broke Curtis's heart. We got Ben outside with the animals and he was okay."

Ben recovered from that seizure, but Curtis never would. From that day on, he committed himself to anything he and his burros could do to help Ben. You only truly bond with your burro when you work together, Curtis believed, so he persuaded the Wanns to join him for outings in the high country. "First time we get there, he takes us five miles straight the hell up Mount Harvard," Brad Wann would later tell me, still a little miffed that he had to walk while Ben got to ride McMurphy, one of Curtis's best racing burros. "Fourteen thousand feet!"

"Why did you name him McMurphy?" Amber asked Curtis.

"After the crazy guy in *One Flew over the Cuckoo's Nest*," Curtis replied, which didn't alarm Amber as much as maybe it should have.

"In my gut, I knew it was okay," Amber would recall. "As soon as Benjamin was on his back, that burro's ears went up. He looked regal. Like he had a purpose: *I have this special package to carry*." Amber was also thrilled to see Ben so happy. "I was this depressed mom at home with an epileptic child, and now I'm thinking, *This could be it. This animal therapy is for real—*"

And that's when McMurphy tripped.

"Craziest thing," Curtis said. "McMurphy goes down in the dirt. I knew what was going to happen next 'cause I've seen it a thousand times. Animal falls, rolls over, comes back up on its feet. That's why so many horse riders break a leg—from the horse rolling over them. But you should have seen McMurphy. He's about to roll and somehow he stops and fights his body back the other way. Some DNA in him to protect that child."

"It's like he suddenly remembered Ben and went '*Whoops*,' and put himself in reverse," Brad said. "I wanted to kiss him."

"You should have," Amber told him. "I did."

For the Wanns, that tumble sealed the deal. It's been four years now and they've been a dedicated burro clan ever since. They show up at every race, three generations strong; Amber's parents even got a pair of mini donkeys so they can hike along behind the grandkids. "Our family doesn't do other sports on the weekend," Brad explained. "We do this." Ben's seizures have disappeared for as long as six months at a time, and his stamina and self-confidence are through the roof.

"It's been so cool to see the joy in Ben's face as he gets off the meds," Curtis reflected. "Burros have brought a measure of sanity to that family."

Like acupuncture and meditation, equine therapy lives in that anecdotal world where plenty of credible people are convinced it works but can't prove why. As a medical approach, it's both older than the Hippocratic oath and newer than Lasik; ancient Greek healers, including Hippocrates himself, prescribed horseback rides as a treatment for chronic pain and emotional maladies, and by World War One it had been adopted by British hospitals to help heal wounded soldiers. Yet in the United States, equine therapy didn't become widespread until the 1990s, when mental health workers began testing its potential with problems that no other treatment seemed to solve.

One of the more remarkable cases involves Rowan Isaacson, "the Horse Boy." In 2008, Rowan was an autistic six-year-old in Texas who would erupt into such violent fits that he couldn't attend school. "Our lives were tantrums," Rowan's father, Rupert, would recall. "Tantrums and the time in between." But whenever Rupert took his son riding, Rowan's mood suddenly calmed. He became so relaxed and focused that Rupert was able to teach him to read while in the saddle. Rupert couldn't explain how it worked, but he had a feeling who might: the wandering horsemen of Mongolia who first domesticated wild ponies more than four thousand years ago. Rowan's parents actually took him on an expedition deep into the Mongolian outback to learn from these master equestrians, and by the time they left, Rowan was a changed boy. Gone were the eruptions, the anxieties, the odd spinning rituals, even his violent opposition to potty training. "Rowan is still autistic—his essence, his many talents, are all tied up with it," Rupert points out. But "he has been healed of the terrible dysfunctions that afflicted him."

Since their return to Texas, the Isaacsons have thrown themselves into their own Texas-based version of a Mongolian nomad camp, creating a center where other challenged kids can do their school lessons on horseback. One of the lead counselors at the Horse Boy Foundation? Rowan himself. Maybe Rowan's transformation shouldn't be so surprising, says Temple Grandin, the celebrity scientist who overcame her own acute autism to become a world authority on animal behavior. People like Rowan and her, Temple explains, think visually. So do animals. That's why "animals, especially for autistic kids, can often be the connecting point between the autistic and the 'normal' human world."

Horse-based therapies have also shown impressive results for issues ranging from combat trauma and sexual abuse to anger management, eating disorders, and addictive behaviors. Few peer-reviewed studies have been produced so far, but the reports that have come in are promising. One survey of veterans struggling with PTSD found that 72 percent showed significant improvement after several weeks of working with horses, while teenagers in custody for at-risk behaviors have shown significantly better impulse control and social skills. And all you need in order to understand why horses are so effective at reducing stress and anxiety, according to researchers at the University of Toledo School of Medicine, is a set of eyeballs: *Look at the size of those things!*

"The horse weighs easily a thousand or more pounds," the researchers point out in their study of adolescent equine psychotherapy, and the animals' sheer size provides "opportunities for riders to explore issues related to vulnerability, power, and control." When you're working with a force of nature that is extraordinarily sensitive to human cues and will act up if approached by someone who is feeling angry or tense, you learn pretty quickly that you'd best keep an eye on your moods and maintain total attention to the moment.

In return, you'll get a comforting surge of endorphins and dopamine—an evolutionary endowment from your hunter-gatherer ancestors, who first relied on horses as hunting partners, escape vehicles, and early-warning security systems. That sense of well-being from animal contact became so firmly encoded in our DNA that, today, we're still instinctively soothed by the sight of whiskers and the feel of a warm pelt.

Yeah, well. Maybe that's half the story, but according to Curtis, it overlooks the special virtue of donkeys. "Everything about burros is rhythmic," he explained. "Their breathing, their movements, all 1-2-3 . . . 1-2-3 . . . Like the perfect waltz partner. They're desert animals, so that's the way it has to be. Keep your rhythm, keep your cool. So Ben comes along and his heart, his breath, all slow to the rhythm."

"*You become what you behold*," he concluded. "How's that for a little cowboy philosophy? It's really Huxley but sounds like something from the range."

With Curtis as their guide, Hal and Mary were already sitting Harrison on a bareback burro when he was still a toddler, letting

Ben Wann prepares for white water and high peaks ahead.

the two of them connect with nothing in between. With Hal leading and Mary spotting behind, they gradually wandered out of the paddock and up into the woods.

"We often sing, recite books, and point out the different types of trees, wildflowers, and animals that we see along the way," Hal noted. "We noticed right away that on the days when Harrison rode, and even on days following a ride, there was a marked improvement in his disposition and behavior, and fewer tantrums."

But Mary was just as impressed by the burros. Keep in mind, these aren't petting zoo ponies; these are powerful animals with fierce self-defense instincts who hate, more than anything, being spooked. Every chromosome in their bodies has been refined by thousands of years of natural selection to stay as far as possible from explosive noise, sudden moves, lunging bodies—Harrison, in other words. An autistic child is a donkey's version of a flashing red light—*Danger Ahead! Beware!*—yet Hal has never had to bring in a burro especially for the Blur. Whoever he happens to be racing with that year always works out just fine.

"I still don't get it," I told Hal and Mary. Even though I saw it for myself when Harrison did his Jimmy "Superfly" Snuka onto the pasture gate and the donkeys didn't blink, the evidence of my own eyes wasn't enough to explain why any animal would suddenly suspend its No. 1 survival instinct just because some kid was having a bad day. They weren't even the same species. I'd been told about Harrison before I got there, and even I was slower than the burros to process what was going on. What were they seeing that I wasn't? How did they grasp in a blink that something that looked and sounded exactly like an attack really wasn't?

"We base our decisions on logic. Theirs are based on sensory perception," Mary explained. "While we're assembling information in our brains, they're relying on a really keen sense of smell and hearing. Their judgment is amazing and lightning fast."

The Blur wasn't the only one benefiting from the Walters' homemade therapy. By nature, Hal is hands-on, independent, a little impatient. He likes to do things his way and do them now, and if something gets in the way, he has a solution: lower his head and go harder. But well before Harrison was born, the burros were teaching him the hard way that bulling ahead would get him nowhere. "Pack burro racing was my training for fatherhood," Hal said frankly. There's no way you're going to alpha-male a burro into doing what you want, so Hal had to take a step back and recondition himself to accept, adapt, and improvise. The burros were leading him to insights he might never have discovered on his own. Like:

#1 *The only thing you need to do is the thing you're doing.*

"Laredo and Boogie taught me the best lesson for dealing with a child like Harrison: you have to have more time than they do. If you rush, you lose."

#2 *Lead from the rear.*

"You're asking a burro to do something very unnatural. Leave his buddies, leave his food and shelter, and run thirty miles into the mountains. You have to make it seem like it's his idea."

#3 *If they do something wrong, it's because you didn't do something right.*

"Their instincts don't always line up with your intentions," Curtis Imrie said. "When that happens, you lick your wounds, get your panther blood back, and go again. Because when they like you, they'll do everything short of open the gate and jump in the trailer. They become your partner. Your buddy. They join you for the adventure."

But just when you think you've got things figured out, that's when the rope is ripped from your hands and you feel like you've been kicked in the chest, and you realize that you—Father of the Year, Mr. Seven-Time World Champion—don't know jack. Good thing your old pal Harper Lee is there to throw a lifeline and remind you of the most important lesson of all:

#4 *What, you think Atticus had it easy?*

"I wanted you to see what real courage is," Hal likes to recite to himself from *To Kill a Mockingbird*. "It's when you know you're licked before you begin, but you begin anyway and you see it through no matter what. You rarely win, but sometimes you do."

Hal peeked out the window at the dipping sun. We'd talked a good while and it was getting late.

"Up for a run?" he asked.

"Okay. Yeah. Well, if you think—" I stammered, mentally scrambling for a way out. Half the reason I'd flown across the country and driven into the Rocky Mountain hinterlands was precisely for this opportunity, to learn at the feet of burro racing's grand master and reigning world champion, but now that the hour had arrived, self-preservation was kicking in with a vengeance. Steph Curry just offered to take me out back and shoot threes, which sounds amazing until Steph Curry actually tosses you the rock and says, "Let's go." If meeting your hero someday is on your bucket list, take it from me: you will never feel more naked, more awkward and useless and stripped to your bare insecurities, than being invited to stand next to your god and show your stuff.

But my chance to back out disappeared the second Hal heard "Okay." He and Harrison immediately headed for the door, and by the time I changed into shorts and followed, they'd already haltered two mammoth burros and were waiting in the driveway. Harrison was dying to come with us, but he agreed to stay home with Mary after I promised I'd teach him my technique for throwing steak knives. Mary, who'd clearly mastered very chill parenting protocols of her own, didn't seem to mind, and scooted Harrison back inside.

"Four or five miles all right?" Hal asked.

"Sounds good." The short course at the World Championship was three times that distance, so if five wasn't all right, I was in trouble.

Hal took Laredo, the seasoned old vet, and handed me Teddy, young and spry but a bit of a numbnut. Hal would lead, figuring Teddy would stick right behind Laredo. But if Teddy got frisky, I was supposed to swing wide to the left and pull back, which would turn Teddy back on himself.

"We'll go easy," Hal promised.

Thirty seconds later, I was alone and empty-handed, watching Teddy disappear in the distance with Hal and Laredo in hot pursuit. I wanted to help, but . . . no way. My head was swimming, my chest was heaving, and if I didn't stop right now, I'd go facedown in the dirt. Holy hell—this altitude was killing me. We'd started off slowly, just as Hal promised, but Teddy was feeling happy and crowded in on Laredo, pushing up the pace. Still, it wasn't bad—until it was suddenly impossible. After a hundred yards, my temples were throbbing and I couldn't catch my breath. I pulled back on Teddy to slow him, but he ignored me. Failsafe plan was turn him by the halter, but when I sprinted a few steps to reach for his head, the world went woozy and my lizard brain took over. *Let go of this rope before you die*, it commanded, and my hand obeyed.

Hal ran Teddy down and waited for me about a quarter mile up the road. I jogged-walked-jogged toward him, trying to save face but feeling the altitude suck the wind out of me every twenty yards or so. Hal's home sits nearly 8,000 feet above sea level, which is far above Mile High Denver but still *waaay* below the summit of the World Championship course at more than 12,000 feet. Jesus Christ! Even racing the World Championship's fifteen-mile short course instead of the twenty-nine-mile long course was going to be brutal. Mika and Zeke had never run that far in their lives, let alone up here in the freaking Himalayas. And how were the donkeys going to handle it?

"Don't worry," Hal called as I wobbled toward him. "You'll get used to it."

I waved my arm, saving my breath. I hadn't even asked him about the creek crossings yet.

20

Zekipedia

The clock was running out, so early one morning in May, we decided to go for broke and take the donkeys into the maze. The time had come to ask Tanya two questions that only she could answer, and the maze was standing between us.

"Ready to rock?" I asked Zeke when he arrived at seven thirty in the morning, barefoot and still a little sleepy-eyed. For once, I already had the donkeys roped and waiting. I'd even brought out his triple espresso so we wouldn't have to doink around inside. "Chuck this back and get your shoes on," I said, handing Zeke the mug. "We've gotta scoot."

Sherman must have sensed that something special was going on, unless he was getting a contact high off Zeke's coffee fumes. He was tanked up on nervous energy, nipping at Flower so she'd chase him and annoying the hell out of Matilda.

"Why's the Wild Thing suddenly so fired up today?" I wondered out loud.

"The Sherman mind is a mystery," Zeke said. "We need donkey Freud to figure him out."

I got between Flower and Sherman, separating them with my body to put an end to Sherman's shenanigans, and led Flower to the road. Mika came out of the house with hand-strap water bottles for me and Zeke, which we both—stupidly—declined. Mika shrugged and kept one for herself, stuffing the zip pouch with some almonds, and then pulled on the visor that she wears when the sun is blazing.

"Everyone good?" I asked. "All righty. Let's see what we've got."

Flower must have picked up on the excitement in the air, because before I even growled out a command, she was off. I struggled to stay with her, wishing I'd either warmed up or chosen a longer rope. Luckily, salvation was just ahead: after three months of passing the same sign for AK's Saw Shop on the gravel road every single day, Flower still hadn't gotten over her conviction that she was doomed to die at the hands of something flat, orange, and inert. She skidded to a halt, giving me a chance to drop my hands to my hips and suck wind. Already I was soaked in sweat.

"Gonna be a hot one," I said, as Mika and Zeke caught up. "But we should be in and out of this thing and on our way home before it gets too toasty."

"Sounds good," said Zeke, who didn't know me any better.

"Are you sure about this?" asked Mika, who did.

I wobbled my hand, so-so style. "Pretty much."

The maze was something I'd been eyeing for a while but never had the nerve to try. Hidden in the woods about three miles from our house, on a steep hill dropping down to the Susquehanna, is an ancient slate quarry dating back to the 1700s. It had been abandoned nearly a century ago and swallowed up by the forest, leaving only a few thin threads of hunting trail. Nearly every time I'd gone in, I'd Blair-Witched around in circles, hunting for the road but constantly finding myself on the edge of the same fifty-foot cliff over the river. Eventually I'd stumble out, exhausted and thorn-scratched and never where I expected to be. I kept at it over the winter, exploring the maze whenever we weren't running with the donkeys, intent on finding a route we could run from end to end. One day, I gambled on a path I'd always avoided because it obviously led straight to the water. I followed it through corkscrew twists and ravines, feeling more lost and regretful by the mile, until suddenly I popped out of the trees. To my confounded amazement, I was on a dirt road that led right to Tanya's front door.

I didn't know what kind of crafty animal logic created that curlycuing roller coaster, but it worked. The trick, I realized, was to ignore my sense of direction and go right anytime my brain screamed "Left!" Whether I could keep that straight while dealing with a trio of donkeys was another story, but that was a big part of the reason we had to try. If we made it to Colorado, we'd be flying blind in those mountains. It would take us at least two days, maybe three, to drive there from Pennsylvania, leaving very little time to scout the terrain. Come race day, our biggest nemesis might not be alpine thin air or thundering creeks, but six different brains making their own survival choices. All of us would be stressed, and weary, and absolutely sure the other five didn't know jack. We already had the odds stacked against us, so if we couldn't pull together when things got awful, it was game over. The maze could be the perfect proving ground.

"*Heeeey-YUP!*" I called to Flower. She glided into a trot, and now that I was loosened up, it was a pleasure to coast along behind her. We clattered to the end of the gravel road, then leaned into the Big One: a mile-long climb up Slate Hill Road that led to the outer edge of the maze. The Big One is a never-ending grind, but Flower handled it like a dream; she flowed uphill so smoothly, I could forget about her and worry about myself. Within a minute, my chest was heaving, so I lowered my head and began counting my steps. *One, two, three . . .*

I made it to ten, then started over, focusing on the numbers to distract myself from the urge to quit. *One, two—*

WHOMP. I slammed into Flower's rump, knocking what little wind I had out of me. She'd stopped so suddenly, I never saw her freeze. I looked around, but I couldn't spot any of her usual roadside triggers. No skid marks, no hanging tree branches, not even her own shadow. Mika and Zeke pulled up, panting.

"You okay?" Mika asked.

"Just some Flower weirdness," I said. I called Flower back to action—*Heeeey-YUP!*—but every time we got a head of steam going, she froze again. Flower was supposed to be our big gun, the ace runner who would pull the rest of us along behind her, and before we even got to the maze, she was screwing things up. Or was she? Pissed off as I was, I couldn't forget Curtis Imrie's words. "Bunch of bull, blaming her," he'd snort. "Figure out your own mistake before you start crying about the burro."

So when we started back up the hill, I quit counting steps. I glanced back and forth between Flower and the road, scanning for her secret phantom, and there it was. I watched her big brown eye roll to the side and look right at . . .

Me.

"You. Big. Stinker," I grunted. *So that's what you're doing*. Flower wasn't reacting to something she saw; she was reacting to something she heard. Every time my breathing got raspy, Flower took it as a signal to stop. And why not? Donkeys are survivors, so if that hill was tough enough to kick the crap out of me, why should she take the same risk? Flower is no dummy. She'd tested my leadership and found two flaws: I didn't have the wind to order her on, and when she tempted me with a chance to rest, I grabbed it. Someone had to step up their game, and it wasn't Flower.

At the top of the hill, we paused to huddle before entering the maze.

"Be careful you don't trash an ankle," I warned. "The first half mile is super scrabbly." Karl Meltzer, the ace ultrarunner who set the Appalachian Trail speed record, once said that of all 2,200 miles he ran across fourteen states, Pennsylvania had the most punishing rocks. They seem to grow straight out of the ground like fangs, jutting up so you feel like you're dancing your way through the mouth of a monster shark. Luckily, the maze was that bad only at the beginning; we faced a half mile or so of shark's teeth before the trail softened into nice, hardpacked dirt.

Flower and I took point. We jogged into the woods, skirting the top of the deep gash left in the hill after hundreds of years of mining it for slate. Nature and neglect had done a magnificent job of healing the place, greening it over so thickly with locust and black walnut and wild pawpaw that I had to backtrack twice before spotting the slit in the trees that marked the entrance to the maze. As I expected, Flower wanted no part of that shadowy tunnel.

"Bring on the 'Tildonkey," I called.

Mika brought up Matilda, and together the two of them sidestepped us and pushed straight into the brush. Sherman hurried to follow, towing Zeke along with him. In a blink, the four of them were swallowed by the forest. Flower watched them vanish, pacing nervously, before deciding a quick death with her friends was better than a long life with me. She plunged through the gap, tapdancing through the minefield of ankle-twisting stones. Flower's agility always astonished me; I kept thinking of her as a big ol' galoot, mostly because of her reticence, but once in motion she's an absolute butterfly. She clattered along without even a glance at the ground, while I stumbled behind her, my eyes darting desperately from rock to rock.

Her courage restored, Flower floated past Sherman and Matilda and slid back into the lead. I kept feeding out rope, letting her barrel along at her own pace, figuring I'd catch up to her when we reached smooth trail ahead. Instead, Flower blasted off when she hit the dirt like she'd been waiting for it all day. I pushed hard to keep up, but something smacked me in the legs. I stumbled, and got thumped again. Sherman and Matilda were crowding in, racing to get past me like a pair of gladiator stallions. All three donkeys were *on fire*, surrounding me in a flying wedge, with Mika and Zeke at full sprint bringing up the rear.

I felt the last inches of my twelve-foot rope sliding through my fingers. Time to get this shit under control, I told myself—but I didn't have the heart to rein those nutballs in. Something about the feel of cool dirt under their hooves seemed to electrify them, just the way it did the first time we led Sherman off the gravel road and he raced up the hill to visit our neighbor's horses. The donkeys were having such a blast, it gave all of us an extra surge of energy. All six of us whipped around turns as the trail dipped and twisted, forgetting to worry about our footing as we were carried along by the excitement of our mini stampede. The donkeys slowed a little as they navigated a tricky stretch covered with jutting roots, giving us all a breather, but the party kicked off again the second we reached the other side.

Flower blazed ahead, storming straight toward a fork in the trail. "Right!" I yelled, waving my hand at Flower's left eye and hoping that in the previous five minutes she'd suddenly learned English. That fork was the same spot where I'd made my *aha!* discovery a few weeks ago that the only way out of the maze was to go right. Flower, of course, went left.

I had only a few inches of rope in my hand, so I couldn't pull her back. Flower plunged happily along, oblivious to the fact that the only thing in her future was a mile-long drop down to the Susquehanna. I lunged, grabbing for an extra handhold of rope, pulling myself hand over hand like a drowning swimmer, gripping a little more each time. When Flower veered around a sapling, I saw my chance. Flower went right, I went left, and the sapling caught the rope between us. It bent like a bow but held, slowing Flower just enough for me to rein her in.

Seconds later, Zeke and Sherman slid in for a landing, both of them stutter-stepping to brake their speed on the steep dirt slope.

"That was—" Zeke gasped. "Awesome!"

I leaned against the sapling while Zeke walked in small circles, both of us belly-breathing hard. Matilda and Mika, the more sensible team, took their time coming down the hill. I should have waved Mika off and saved her the long hike back up, but I was too gassed to think clearly. Sunlight was stabbing through the trees, heating the maze into an open-air oven. Zeke and I pulled off our shirts and knotted them around our heads like bandannas, letting the long ends drape behind to protect our necks. We'd been out for only an hour, but I was already parched and my stomach was rumbling from skipping breakfast before the run.

Mika took a swig from her water bottle and passed it around. She counted out her almonds and divvied them up with Zeke and me, kindly neglecting to point out that the two rockheads who'd declined their own bottles back at the house deserved a dope-slap instead of a snack.

"I won't make that mistake again," I promised.

But privately, I knew that wasn't good enough. A friend of mine at MIT's business school likes to remind his graduate students that problems come in two flavors: "Stubbed your toe" versus "Lost control of the chain saw." You need to determine the degree of damage, in other words, before you decide the degree of response. For corporations, that's an excellent battle plan, but for experienced burro racers, the math doesn't add up that way. You can be running with your burro in a beautiful groove, both of you perfectly in sync and feeling awesome, when a little wrinkle bunches in your sock. It would be stupid to ruin your flow over something so teensy, so you ignore it and muscle through . . . until the blister rips open, the blood seeps through your shoe, and your limping gait upsets your burro and ends your race, leaving you lame and stranded at 13,000 feet. With donkeys, there's no such thing as a tiny glitch. Every stubbed toe is a gathering storm.

We can't worry about just the donkeys anymore, I told myself as we munched the last of the almonds, fueling up for our next push through the maze. Deep inside, Sherman and Matilda and Flower had such a core of strength and loyalty that they would follow us into nearly any adventure. But it was up to us to lead them back out again, and I wasn't sure I was up to the job. So far in the maze, I'd dropped one chain saw after another. I'd flamed out on the big climb, lost control on the trail, and gotten caught in the heat with no water. In the Rockies, any one of those flubs could be a disaster, and I'd rung up all three in a single hour. We couldn't just train the donkeys anymore; no matter how well prepared they were, we had to keep one step ahead. We had to be a little stronger. Always.

"Time to climb Everest," I said, and we began trudging back up the hill.

The donkeys took it easy on us, letting us hike comfortably beside them until we reached the top and eased into a jog. This go-round, I forced myself to stay tight by Flower's side, keeping enough coiled rope in my fist to steer her through the forks in the trail. A few times, I was so focused on Flower that I missed the turn and had to backtrack, but no one seemed to care. The snacks and water had revived us. We weren't lost in a maze anymore; we were running wild in the woods and having a blast. Before we knew it, we crashed out of the trees and onto the dirt road to Tanya's house.

"That's my girls!" Tanya hooted when she saw Flower and Matilda trotting down her driveway. "Oh my *god*! Is that Sherman? Is that the same Shermie Germie?"

Tanya was overjoyed. She crooned over Flower and Matilda, massaging their ears until their bottom lips drooped with pleasure. She hadn't seen Sherman in a few months, and she couldn't get over how much he'd changed. "He looks terrific," she said, then made sure by squatting to check his hooves. She nodded her approval. "Even his feet are looking better."

After Tanya and her husband split up and we lost Sherman's hacksaw hoof savior, we'd put out a distress signal on the Amish word-of-mouth hotline. It turned out our neighbor over the hill, old AK, had a son a few miles away who could do the job. Whenever Matilda needed a trim, she'd hop into the back of our minivan, poking her head into the front seat so she could enjoy the view as I drove. Sherman and Flower were too big for the van, so we'd run them cross-country through the hayfields, waving to the Amish kids who clustered along the fence to watch as we cut behind the baseball field next to their schoolhouse. Sherman must have loved getting his hooves trimmed, because on the way back, he'd prance like a show pony.

"You think he's looking pretty good?" I asked Tanya.

"God, yeah. Remember that bloating along here?" she said, swatting his flanks. "All muscle now. You'd never know it was the same Germie. You can see in his eyes how happy he is."

"Sooo. . . ." I began, preparing to hit Tanya with the first of my two do-or-die questions. "The race is in two months. We've really got to ramp up training. Is it fair to Sherman to work him that hard?"

Tanya chewed it over. "How's it been going so far?"

"Some days great, some days not," I told her. "They'll run four, five miles, no problem. Then I'll try to go longer and they'll just shut down. Remember the Amish Full Moon Run? Same thing. We'll be cruising, then suddenly they blow a fuse."

"That's up here," she said, tapping Sherman on the head. "Remember what I said about a sense of purpose? You taught them the wrong job."

Somewhere along the line, I'd accidentally strayed from Tanya's number one rule: Make the donkeys think everything is their idea. The best jobs, she'd explained to me, are like the best dinners. You tuck into them because you're hungry, and when you finish, you feel strong, happy, and fulfilled. I'd thought I was doing the right thing by easing Sherman into new challenges, spoon-feeding him along, but that was like getting the same dry meat loaf every night. It was time to spice up the menu: it was time to put our donkeys on the Bomb Dog Diet.

I'd first learned about it from Alexandra Horowitz, the PhD psychologist who specializes in the canine mind. Horowitz has spent a lot of time observing bomb- and drug-detection dogs, because they're among the most rigorously educated animals in the world. Sniffer dogs have to locate fantastically small traces of explosives—

we're talking *a trillionth of a gram*—even when camouflaged by scents as overpowering as coffee grounds or Vanillaroma car air fresheners. Success means saving lives and averting massacres, but it also means ignoring every other instinct and distraction and focusing so intently on the task at hand that, in the midst of havoc, these dogs are able to home in on a hidden chemical that no other machine or creature alive could detect. And do you know what they get for a reward? Not a juicy bone. Not even a handful of treats.

An old, chewed-up tennis ball.

Detection dogs go nuts over a toy like this, because it's something they can chase and hunt and track down—just like they do on the job. Their occupation is so gratifying, in other words, that the best thank-you you can give them for doing it well is to let them do it some more. Could donkeys feel the same way?

"All creatures have a biological imperative: *The sun is up, so how do I fill my day?*" Horowitz told me one evening when we met before a screening of *Isle of Dogs*. She was there to talk canines, but she was gracious enough to take my questions about donkeys. "By domesticating animals, we can remove their evolutionary purpose," she explained, "and that can lead to problems"—no surprise if you've ever arrived home to discover your springer spaniel has stalked and eaten your dress shoes.

This is easy to grasp if you remember that we humans are just animals in clothes. "The differences between us are trivial compared to the similarities," Horowitz pointed out. So if you and I are hungry for a challenge, for some task that feels urgent and perfectly suited to our skills, why wouldn't every other creature? It's tricky, though; we humans made things a lot more complicated when we took over job distribution for most of the planet and stopped letting animals choose for themselves. Dog shelters are packed with living proof of how often people get it wrong, but it's not that hard to do it right.

"The best situation," Horowitz advised me, "is to find a coordination of purposes." A job that matches their natural drive. Dinner that feels like dessert.

. . .

"You've got to retrain them," Tanya explained to me, Zeke, and Mika in her driveway. "You need to mix things up, let them know every day is different but one thing is the same: their job is to stay by your side until you decide to stop." Somewhere in Sherman's brain, a factory whistle screamed for quitting time as soon as he put in a solid hour of running. Instead, he should be having such a blast that his reward for running with us should be . . . the joy of running with us. One part of Sherman's brain would always be calling him back to Lawrence and that cozy little barn, but my task was to make the workouts so much fun and so fulfilling that he forgot all about the barn for a while.

"So for the race," Tanya asked. "What's the starting line like?"

"Oh god. Mayhem."

"Okay, you have to do mayhem. How about the finish line?"

"Worse. Spectators are screaming. And oh yeah—sometimes you crash into the outhouse race." I'd forgotten about this until just now. While waiting for runners to return, the Pack Burro World Championship likes to entertain the crowd with a shithouse sprint: four-person teams push wooden outhouses on wheels down the middle of Main Street, so many of them that heats, semifinals, and finals are needed to crown a champion. It wasn't unusual for a burro racer to finally struggle home after twenty-nine miles of mountain running, only to lose her burro in the last hundred yards because it got spooked by a Ben Hur gladiator derby of giant wooden crappers.

"Perfect!" Tanya said. "That'll get them rockin' and rollin'. Keep throwing crazy at them. The harder you make the training, the more they'll respond." I drank it all in, feeling more optimistic by the second. It was so great to be around Tanya again. I'd forgotten how uplifting she can be, even when she's pushing me to cowboy up and get to work. Her mind seems to operate in terms of dreams and adventure, even though she's always blunt and honest. No wonder her animals love her. I wasn't 100 percent sure I understood what she meant by "throwing crazy," but whatever, it sounded cool. We'd figure it out.

"They're great donkeys," Tanya concluded. "They won't let you down. Look how great they did for you today. That's really good on a scorcher like this."

"Actually," Zeke piped up, "these conditions are ideal for donkeys." To my horrified amazement, he went on to explain to Tanya—*to Tanya*, who grew up in the saddle, who majored in equine science and single-handedly kept Sherman alive—yes, to that Tanya, Zeke decided it was appropriate to deliver a lecture on the descent of the modern donkey from the wild African ass, pointing out that really, as desert animals, a hot day like today was right up their evolutionary alley.

I forgave Tanya in advance for telling Zeke to blow it out his evolutionary ass. Instead, she swatted him on the back. "Where'd you find this guy?" she asked with a laugh. "Thanks, Zekipedia. Desert animals or not, they're going to get fussy. You guys better get moving."

"One last thing," I ventured, already regretting the last question I had to ask. I knew Tanya was still struggling to make ends meet and keep Christmas Wish Farm afloat after her marriage had suddenly ended. We'd barely seen her over the past few months, and when we did, it was usually because she was zooming by on another last-minute driving job for an Amish farmer. I was about to put Tanya in a shitty spot, but with only two months to go, I didn't see any way around it.

"What do you think about hauling us to Colorado?" I said, as I cringed inside. "Any chance that's still on the table?"

"Are you serious?" Tanya retorted. "I wouldn't—"

"I'm sorry," I quickly apologized. "I just wanted—"

"I wouldn't miss it!" Tanya finished. All winter long, the one bright spot she'd been looking forward to was the day she could watch Shermie and me push our way into the ranks of world-class racing burros and try our luck high in the Rockies. She had already calculated how much time she could take off and who would watch out for her horses and Dobermans while she was away. I should have known better. Did I really think for one dang second that some wayward ex-husband was enough to stop Sherman's fairy godmother from getting him to the ball?

"How big a trailer are we talking?" Tanya asked. "Is it just Sherman, or all three of you now?"

This time, Zekipedia kept quiet. His eyes were pinned on the ground, suddenly very intent on interpreting the gravel. He didn't say a word, but everything he felt was written on his face. Mika and I glanced at each other. We hadn't discussed it yet, but at some point in the maze, we'd made up our minds.

"Absolutely," I said. "We're a team."

21

Go Barb Dolan on That Burro!

A few weeks later, I was in front of the house pounding fence posts for a new donkey pasture when a car whipped into the driveway. A woman with a frantic look in her eye waved me over.

"Chill the eff out," I muttered, irked by the interruption but kind of pleased that I'd managed to stick with my recent F-bomb ban. "Never saw a freaking goat in the road before?" I added, blowing the streak because I've never actually said "freaking" in my life. I dropped the digging iron and headed over to the car. "Hang on, I'll grab him," I called, looking around for Lawrence before realizing he was traipsing along behind me. Whoever this stranger was, whatever upset her, this time Lawrence had an alibi.

"You're Chris?" the woman asked.

"Yeaaah?" I answered warily.

"Paul sent me over."

"Paul?"

"Tanya's friend. She's been in an accident. It's serious."

At that moment, an ambulance was speeding Tanya to Lancaster General Hospital. "She was helping me train my new carriage horse," the stranger told me, "when the horse spooked and went berserk, flipping the carriage and dragging her." The woman managed to grab the reins and stop the horse, but not before Tanya had been badly battered. Paramedics stopped the bleeding and immobilized her on a backboard, fearing she might have suffered a spinal injury. "I'm Shelley," the woman finally mentioned. She had called Paul, who told her to alert me before heading to the hospital herself.

I ran to the house for my phone and heard it ringing before I got there. Somehow, my neighbor Amos had already gotten word of the accident and was looking for a ride to the emergency room. "Sit tight," I told him. Mika happened to be in Lancaster that morning, so she could get to the hospital first and find out what was going on. I knew that the second Tanya opened her eyes, she'd be fretting about her animals, so she might get more peace of mind if she knew Amos and I were looking after her farm rather than crowding around her hospital bed.

I called Mika, who instantly sped to the ER. Tanya was conscious, Mika told me, but in a lot of pain. She had suffered a cracked skull, two broken vertebrae, and a badly gashed knee. Thankfully, the doctors were confident that Tanya would recover but warned that she was facing a long road back. She'd be in a body cast for at least two months before she could even begin to regain mobility.

Amos and I drove over to Christmas Wish Farm and found Tanya's neighbors already at work. They'd fed and watered the Dobermans and horses, mucked out the barn, and even dragged the trashed carriage out of the paddock. Tanya's place would be well cared for, they assured us, and her house would be clean and ready for her return. Once she was home, the Amish community would cook meals for her and mobilize a chicken potpie fund-raiser to contribute to her medical bills. Tanya might live in the middle of nowhere, but she wasn't alone.

"How about you?" Amos asked as we headed home. "What are you going to do?"

I knew exactly what he meant but pretended I didn't. Amos had helped us with Sherman from the day he arrived, even volunteering his younger brother to care for Sherman's hooves when Tanya's husband suddenly exited, so he'd witnessed firsthand how much we relied on Tanya and stumbled when she wasn't around. Still, it felt monstrous to be already worrying about myself so soon after her accident—until I remembered that this was the way Amish farmers deal with tough breaks: As soon as they see a problem, they attack it. What other choice did they have? If you don't have the luxury of tapping a few numbers on a phone to save your ass when you've ignored the drip under the sink or forgotten to shop for dinner, you learn you can't put things off and hope they magically get better on their own. You've got to square up to the facts, and that's why Amos was urging me to face this new reality right away: Tanya wasn't coming back. As of that moment we were on our own, and we didn't have a moment to spare. Until then, I'd assumed Tanya would be back to help us through the final weeks and join us at the starting line, but the bottom had just dropped out of that plan. We wouldn't even see the starting line if I didn't figure out—and fast—how to haul three donkeys to Colorado without Tanya, her truck, or her trailer.

"You know anyone who can drive us?" I asked.

Amos shook his head. "I'm not even sure who to ask."

I dropped Amos off and drove home slowly, thinking hard. By the time I reached the driveway, I had it. God, the solution was so obvious. Why did we need a driver when we already had three? Mika, Zeke, and I would just haul the trailer ourselves. I'd ruled out that possibility ever since Tanya once described to me how dangerous and difficult it is to get donkeys back into the trailer once you've stopped on the road to walk and water them, but then I had the brainstorm that cracked the problem wide open: we just wouldn't stop. If the three of us split the driving, we could bomb all the way to Colorado in one shot, taking turns at the wheel and sleeping in the backseat between shifts. What would it take—about a day, day and a half? Easy.

I sat in the driveway for a while, stress-testing the scenario before going into the house to present it to Mika. Obviously, we'd have to take it slow for the first few hundred miles, since none of us had ever driven a trailer before, let alone a trailer full of live animals prone to kicking. We'd need to borrow Tanya's Durango, which would demand a little adjustment as well: it was finicky and had a quarter-million miles on the engine, and I think a stick. Did Zeke even drive stick? Her trailer was just as ancient; the tires looked bald and she'd had to nail down plywood to cover the holes in the floor. But hang on, we couldn't be using that rig anyway; it held only two donkeys. Which meant . . .

Screams, sirens, the agonized shrieks of dying donkeys. My mind was suddenly filled with the image of a horrific highway accident: the Durango shattered and strewn in pieces across the asphalt, the trailer ripped apart like the old tin can it was, Sherman and Flower and Matilda mangled and bleeding, while the rest of us . . . Nope, not even going there. That's where my imagination shut down, refusing to even envision what would happen to Zeke and Mika when we crashed—which we absolutely would. Put three rookies in a decrepit Durango that's already ticked past a quarter-million miles and send them westward pulling a 5,000-pound stock trailer, and you know that somewhere—maybe at midnight in the Blue Ridge Mountains, possibly during rush hour in downtown Indianapolis, most likely on one of the hairpin passes high in the Rockies—that little gamble is ending in a morgue. I'm usually more stubborn than Sherman when it comes to letting go of a bad idea, but that image of blood on the road made me dump this plan for good before I even got out of the truck.

Amos was right about one thing, I realized as I went inside the house: fretting about the wrong problem was leading me toward trouble. Right now, there was nothing I could do about our transportation fiasco except send out a call for help and be patient, same as I did when we needed a new hoof healer. In the meantime, there was another situation that needed immediate attention. I checked the time. Nearly eight at night. Still early in Wyoming.

Time to make the call.

. . .

"Hey, Speedy!" a cheery voice on the phone greeted me.

No one except Eric Orton could give me that nickname without it sounding like a put-down. He knows better than anyone that nothing can make me fast and nobody can make me care, which in my eyes makes him the perfect coach. I met Eric during my previous life as a magazine journalist, when *Men's Journal* assigned me to profile this innovative fitness trainer from Jackson Hole who was among the first to use natural-movement techniques to make his athletes stronger, calmer, and more resistant to injury. Back then, I was overweight and aging, and effectively banned from running by doctors who said my Shrek-like body couldn't withstand the impact of "all that pounding."

Horseshit, Eric snorted, and then he proved it: within nine months, he'd ripped me apart and put me back together again, building me up from scratch until I could run a fifty-mile race with Tarahumara tribespeople across Mexico's Copper Canyon. Coach E gave me a gift that changed my life. He not only made *Born to Run* possible; he sent me away from that adventure feeling absolutely unbreakable, as if I could step out the door every day and run as far and as hard as I felt like without worrying about getting hurt—which I've pretty much been doing ever since. Occasionally I'll pull a muscle or get sloppy about technique, but all I have to do is reboot with the lessons Eric taught me and I'm right back in action.

"How's it been going?" Eric asked, when I called him after Tanya's accident. "You excited?"

"You wouldn't believe the crazy drama. We could use some serious help."

"Nice," Eric said. "Way to keep things spicy. So what's going on?"

Eric, naturally, was one of the first calls I'd made last fall when I'd first begun toying with the idea of a burro race. I wanted someone I trusted to tell me, flat out, if I could handle that kind of a beating. At age fifty-four, I would be preparing for an uphill run in thin air on the highest drivable road in North America. The Fairplay route begins at 10,000 feet and climbs to more than 12,000, nearly half the height of Mount Everest. I hadn't trained for something that hard in a decade, not since the last time Eric helped me reverse years of inactivity, and I knew he would give it to me straight. I'd learned that lesson down in the Copper Canyon when Eric and I crossed paths in the middle of the race as he was descending a big hill I was about to climb. I expected him to cheer me on. Instead, he dropped this cheery basket of fun on me: "Brace yourself. It's a lot harder than you think." He totally knee-chopped me—until I began running again, and found as I leaned into the hill that the hard facts were a lot more helpful than a happy lie.

So when I asked Eric at the beginning of my burro training to handicap my chances, I steeled myself for another punch in the gut. Instead, he got all weird and karmic.

"So eerie," he said. "You know how eerie this is?"

"Not a clue. What are you talking about?"

"Dude, it's ten years to the month since we trained for the Copper Canyon race. It could be the same week. It might be the same *day*," Eric said.

Fair enough. The timing was pretty cosmic. But from a more earthly perspective, it also drove home the fact that I hadn't trained hard in a *looooong* time. "What do you think?" I'd asked. "Can we do it again?"

In that very first conversation, Eric had gotten right down to business. "Okay, we've got a lot of work to do. But yeah, you can handle it. You've got a great motor, and when you're fired up about something, you see it through." Eric quickly launched me on a workout schedule that gradually increased my mileage while fine-tuning my running form to prevent over-use injuries. I followed Eric's formula whenever we weren't training with the donkeys, and week by week, I began feeling stronger, faster, and lighter on my feet.

But then things got complicated. Eric thought he was signing on to work with one man and his sick donkey, and before he knew it, I'd added my hula dancer wife to the team, taken a depressed college kid under our wing, tripled the number of donkeys, realized one of them was too fast for me, discovered we had to cross a raging torrent, and lost our trailer driver and head trainer to a buggy crash.

"We're getting down to the crunch," I told Eric. "We've got a lot of problems and not a lot of time."

Eric corrected me: "Flip it around," he said. "Remember what I told you with the Copper Canyon race? You don't start with today and aim toward your goal. You start with the goal, and aim back toward today. Do it like that, and you'll always find a way."

Eric was right; I'd forgotten we'd had the same conversation ten years ago, back when he'd also had to talk me off the ledge when I was convinced that nine months was far too little time for anyone, let alone an out-of-shape doughboy like me, to go from zero miles a week to fifty miles in a day. It's not magic, Eric had taught me. It's not even toughness or luck; it's just math. Spot your finish line, count the steps to get there, and take them one at a time.

"But you've got a point," he went on now, not missing a beat as he shifted from pep talk to whiteboarding. "This is a lot trickier than just running up and down a mountain." The way he saw it, we were fighting three tugs-of-war at the same time. We had to prepare for high altitude, but at sea level. We had to train with the donkeys, but stay one step ahead of them. Mika, Zeke, and I had different ages, genders, and experience levels, but we had to run together as a team. Eric loved the challenge, and promised he would dive right in and get back to me with a battle plan.

"By the way," he added, "have you told Mika she can win?"

As usual, Eric had done his homework. He knew a curious fact about pack burro racing: some of the greatest showdowns have been duels between men and women, and as likely as not, it ain't the guy who comes out ahead. Once that shotgun blasts and the mad stampede disappears into the mountains, it's anyone's guess whether a man or a woman will be the first to return.

In 2011, for instance, forty-year-old Karen Thorpe went up

against three of the fastest burromen alive—Tom Sobal, Justin Mock, and Jim Anderegg—and outran them all to win the 63rd World Championship. Louise Kuehster has been a top contender in the short courses since she was a teenager, and Barb Dolan triumphed in one of the sport's most thrilling sprint finishes when she and her trusty partner, Chugs, fought off Bobby Lewis and Wellstone in a furious final charge to win the 2004 Buena Vista Gold Rush twelve-miler. Burro racing is one of those rare, beautiful, come one/come all events that doesn't care a hoot about the size of the dog in the fight, or even the size of the fight in the dog: muscle and testosterone are no match for stamina, patience, and respect for your teammate.

"Whatever you're hiding inside, a burro will sniff it out," Curtis Imrie always says. "You can't be a bully or a blowhard, and if that sounds more like one gender to you than another, you'll understand why men can struggle at this sport and women excel."

I learned that lesson by accident while I was visiting Hal Walter in Colorado. That same weekend, Hal was hosting his annual Hardscrabble Mountain Run, a 10K/5K trail race across the nearby foothills with no burros involved. I wasn't exactly psyched to run after I'd been brought to my knees by the thin air the day before, but I couldn't blow off Hal's race after he'd been so hospitable. Surprisingly, I got off to a decent start—or so I thought, until I was halfway up the first big climb and some ponytailed kid blasted past me like she'd been launched from a cannon. She disappeared so far into the distance, I didn't see her again until Hal was handing out the awards.

"C'mon up, Lynzi," Hal shouted, calling out first place in the nineteen-and-under division. "Helluva burro racer too, folks. Lynzi was only fourteen last year when she finished fourth in the World Championship."

Fourteen and finished fourth . . .

Hang on. Did I hear that right? Was Hal saying some ninth-grade girl not only handled the same kind of beast that had humbled me the day before during my training run with Hal, but did it so flawlessly that an army of seasoned racers couldn't catch her? All weekend, I'd been soaking up a greatest hits of donkey disasters, cautionary tales of epic crashes and nasty bites and well-meaning McMurphys who trip over their own hooves with epileptic tweens on their backs, and just when I was sinking into a dark pit of doubt, from out of nowhere this human ray of hope suddenly appears . . .

And just as quickly disappears.

By the time I got to the awards area to ask Lynzi her secret, she'd already picked up her hardware and left. I pushed through the crowd, searching for her constantly vanishing ponytail, and spotted her just as she was getting into her mom's car. If I'd looked left instead of right, they might have been gone before I got there. That's how close I came to missing one of the greatest stories of death, love, and triumph I've ever heard.

Plus, I never would have learned the Trash Wrapper Trick.

"People who knew Lynzi as a child can't believe she's even alive," Lynzi's mom, Kelly Doke, told me. She'd invited me over to the woodland home that she and her husband, Ryan, still have to pinch themselves to believe they own. The fight to save their little girl's life was so devastating that the family ended up penniless and displaced, losing their house and everything they owned. Only now, fifteen years later, are they finding their footing again.

On the day Lynzi was born, Kelly was opening the door of their pickup when a blinding pain across her belly doubled her over. Ryan helped her into the truck and floored it, screeching up to the hospital just in time for Lynzi to emerge eight minutes later. Kelly is a nurse, and even though she was still woozy from her own ordeal, she immediately sensed something was wrong. She could hear the nurses whispering, but she couldn't hear her baby. She called out for answers, insisting on being told what was going on, until one of the nurses finally told her: Lynzi was drowning. The pain Kelly had felt in the driveway was her placenta tearing away, allowing fluid to seep into Lynzi's lungs. Lynzi was also six weeks premature, leaving her heart and lungs underdeveloped and vulnerable to infection.

After a few tense minutes, the nurses cleared the congestion and Lynzi began to breathe steadily. The Dokes took her home, but four weeks later, Kelly brought her back to the hospital. Lynzi wasn't eating well, her breathing was raspy, and she seemed listless. "No, she's fine," her pediatrician assured her. "You're a nurse, so you're overthinking this." Kelly had worked with that doctor and trusted him, but something in her gut told her he was wrong. Ordinarily, her husband is the peacemaker of the family, a quiet presence who's clearly enchanted by the way Kelly runs the show. Kelly wouldn't even date him at first; she was so focused on working her way through college as a veterinary technician that she let Ryan spend time with her only if he rode along on her emergency calls for cow C-sections. When Ryan tried courting her with roses for Valentine's Day, Kelly was appalled at the cost and insisted he return them and take her to Taco Bell instead. After they were married, Ryan was happy to let Kelly call the shots with their kids, especially when it came to medical care.

But not this time.

To hell with them, Ryan said when Kelly got him on the phone from the doctor's office. *If they're not listening to you, take her to Oklahoma.* Kelly strapped her four-week-old baby back into the car and drove more than a hundred miles across the state line to another hospital. There, physicians complimented her intuition: Yes, Lynzi tested positive for a respiratory virus. But it was very mild, so Kelly was free to take her home.

"No," Kelly said. "This doesn't feel right."

"There's only one little spot on her lungs," the doctors pointed out.

"Please check again," Kelly insisted. They did. And when that test came out the same, she asked again. And again—and that's when all hell broke loose. In a matter of hours, the single infected spot had erupted, covering Lynzi's lungs with a ghostly white film that started choking her to death. A doctor slapped the Code

Blue button, and medics swarmed into the room. Within minutes, Lynzi and Kelly were on a Flight for Life chopper, en route to a specialty unit in Texas. As soon as they touched down, an ER doctor took one look at Lynzi and leaped onto her gurney, giving her chest compressions as she was wheeled in for emergency transfusions to replace her oxygen-starved blood. She was briefly taken off the respirator to see if she could breathe. Moments later, doctors were threading a tube down her throat: all of Lynzi's systems were shutting down.

Kelly stared, numb with shock. One thought penetrated her panicky haze: If she had obeyed the pediatrician in Oklahoma, her baby would have died in the backseat of her car on the drive home.

Day by day, Lynzi hung by a thread. The doctors warned Kelly and Ryan to prepare for the worst. Nine out of ten babies in this predicament—premature, undersized, heart and lungs under lethal pressure—would never recover. Kelly wasn't even able to hold her suffering baby; whenever Lynzi sensed her mother was near, she would go into spasms of excitement and begin choking. For weeks, Kelly had to gaze through thick glass as her baby struggled to stay alive. She watched in agony every day as a respiratory specialist whacked Lynzi with such violence that the baby was jolted across the bed. "Believe me, it's good for her," a nurse assured Kelly. "He's breaking up congestion. If she doesn't fly in the air, he's not doing his job." Out in the hall, Ryan was dealing with his own nightmare: every few days, he'd see a gurney roll past covered in sheets, carrying another infant who had died of the same infection that Lynzi had.

But bit by bit, Lynzi got stronger. After two months, she had defied the odds and was stable enough to go home. The Dokes returned to Colorado, but found that the life they'd left behind was gone. As a farmer, Ryan paid for his own health insurance, and he was stunned to discover that it covered so little of Lynzi's care that the Dokes now owed the hospitals more than $1.5 million. Lynzi would continue to need expensive treatments and monitoring; just to safeguard her against another infection, for example, each of the

Dokes' two older daughters needed $40,000 worth of vaccinations. The Dokes had to sell their home and everything they owned and then begin the struggle to rebuild. Ryan left farming to work double shifts through the bitter Colorado winters as a telephone lineman, earning better health benefits but getting home so late each night that he barely had time to kiss Kelly in the doorway as she left for her graveyard shift as overnight nurse at the federal prison.

Kelly could deal with her own exhaustion. It was Lynzi's that began to worry her. By the time Lynzi was four, Kelly kept finding her asleep in odd corners of the house. "She'd find any small space—under a desk, behind a laundry basket—and tuck in there for a nap," Kelly said. "She'd sleep in awkward positions, balled up in like a fetal position." She called her doctor, who told them to bring Lynzi in immediately. Testing confirmed his fears: Lynzi's heart was failing. The high altitude of the Rockies was too much strain for her damaged respiratory system, causing her frail heart to work overtime. "If you stay here," he told the Dokes, "your daughter will die."

Once again, Ryan and Kelly abandoned everything they had and started over. They packed up their three kids, said good-bye to both their families and the mountains they loved, and moved into a basement apartment in the flatlands of Missouri. Ryan and Kelly found work, and over time, the family began to recover. So did Lynzi. When she reached middle school, she began jogging with the cross-country team, hoping it might strengthen her weak lungs. The experiment was so successful that within three years, Lynzi skyrocketed from a recovering sick girl into one of the top track prospects in the district. The Dokes were overjoyed, because Lynzi's success also meant something else: she was now strong enough for the family to return to Colorado.

On their first week back home after nearly twelve years, Kelly brought Lynzi to the dentist for a checkup, and was happy to discover the hygienist was none other than her old pal Barb Dolan. Kelly and Barb had both worked in the same prison medical unit, and they hadn't seen each other since the Dokes had made their forced migration to Missouri.

"Wow, how long have you been back?" Barb asked.

Kelly answered that question with one of her own, and it showed better than anything else just how out of touch she'd been: "Any chance you still have your burros?" she asked Barb.

Any chance? Is there any chance the sun is still rising and the earth is still round? Because you'd have better luck betting on those two coming to an end than on Barb's being done with donkeys. True, Barb did retire from the sport. She hung up her halters and called it quits—before roaring back with a vengeance two years later and beating every man and woman in the field in the Leadville twenty-one-mile race.

Wait—that's not technically correct. No one would ever associate Barb Dolan with the words "roar" or "vengeance." Mention her name to anyone who knows her, and you'll trigger an instant involuntary response ("She's such a sweetheart!") followed by a barrage of stories to prove the point. When Caballo Blanco, the grouchy wanderer from *Born to Run*, turned up in Leadville with the flea-bitten Mexican stray he'd adopted, guess who opened her door and her fridge, even though she barely knew him? And continued to offer him clean sheets and hot meals from the day they met until the day he died?

"Barb Dolan is the sweetest, quietest, most warmhearted person on the planet," says Roger Pedretti, a fellow burro racer from Wisconsin. "Until the gun goes off. Then you'd best be out of her way. We have a saying when someone is slowing down: *Go Barb Dolan on that burro!*"

Barb and her identical twin sister grew up in Boulder, where they both became U.S. National Team cyclists. But Barb also enjoyed running, and more and more she found herself leaving the bike behind so she could scramble the dirt trails. She was so torn between her two passions that in 1994, she created a whole new performance standard: she became the first person ever to complete both the Leadville 100 Trail Run *and* the Leadville 100-Mile Mountain Bike Race in the same month. Today, that achievement is called the Leadman Challenge, but it should really be the Double Dolan, or at the very least the Lead*woman*, because no one attempted it until Barb showed it was possible.

"You'd kick butt with my Dinky," one of her trail-running buddies commented, which made no sense to Barb until he filled her in on the basics of burro racing. Even though Boulder is a hotbed of endurance athletes and only a few hours from Fairplay, the two are separated by the iron curtain dividing Carhartt from kombucha. Boulder is Prana and JUULs; Fairplay is Wranglers and Copenhagen. His Dinky, Jim Feistner explained to Barb, was a fast and extremely well-trained racing burro. But unfortunately, Dinky couldn't compete in the upcoming World Championship because Jim's knee was dodgy. Unless Barb wanted to step in?

"It sounds crazy, but yeah, I'll give it a try," Barb said. "What do I do?"

"Simple," Jim said. "Just hold on to the rope."

Easier said. "What blew me away was the *speed* those donkeys take off!" Barb recalls. "I was hanging on to Dinky for dear life. Luckily, Mary Walter took pity. She yelled, 'Wrap the rope around your butt! Slow him down!' Oh my god, I have never run so hard in my life. When I finished, I think I was crying. I swore I would never do that again."

But as the soreness faded, Barb kept ticking over the race in her mind. What made Mary Walter so much better than she was? Barb was stronger, faster, and more experienced as a competitor—she was an Olympic-caliber cyclist, for Pete's sake—and her burro was every bit as speedy as Mary's. So why did Mary soar through the mountains while Barb suffered? As an athletic challenge, burro racing was a lot trickier than it looked.

"I've got to figure this out," Barb decided. Once Barb commits, she has no second gear. She went full throttle from the get-go, even hitching a trailer to her truck and driving all the way to South Dakota after she saw an interesting-looking donkey for sale in a magazine. The donkey she wanted was gone by the time she got there, but Barb noticed an even bigger one in the pasture. "How about that fella?" she asked. "Mind if I take him for a run?" The owner had no idea what she was talking about, but handed Barb a halter and wished her luck.

"I roped him up and he took off like a scalded-ass ape," Barb would later say. "Maybe he liked stretching his legs, maybe he was wondering what this lady was doing behind him. But holy goodness, that guy was fast!"

Not to mention, wild. Barb brought Chugs home and dedicated herself to mastering the art of burro training. For the first three months, Barb put in a full day of work and then spent every evening walking Chugs around her big rectangular paddock. She was trying to let Chugs know that his job was to go point and set pace, but Chugs had his own approach, which often involved spinning back and bowling Barb over the second her mind strayed. "He'd always keep an eye on me to see when I wasn't looking," Barb says. "Once, he got me so hard he knocked me out." Still, Chugs was magnificent on the trails—as long as he was *on* the trails. They'd be rolling through the woods in perfect sync, Barb says, "and then *BOOM!*, he'd go tearing into the trees. I'd get him out and we'd start again, and *BOOM!*, he's dragging me off the other way."

Animal-human partnerships don't get much rockier than a teammate who's constantly threatening you with concussion or escape, but Barb was willing to blame part of it on her own inexperience. She would pepper Mary Walter and Sue Conroe with questions, and the two veterans would invite Barb to join them for all-day training sessions high in the Rockies. Barb and Chugs began to figure each other out, and the better they communicated, the faster they ran. By the end of three years, Barb had been through an apprenticeship that would have put a Shaolin monk to shame. But when she came out the other side, there was nothing a donkey could throw at her that she couldn't handle.

The gun had gone off. Time to get out of her way.

"That's when I became a force to be reckoned with," says Barb. That's not a brag; that's an undersell. For the next two decades, this gentle dental hygienist was one of the most ferocious burro racers

Barb sprints to another World Championship.

on the planet. Barb won the women's Triple Crown an astonishing thirteen times, including ten years in a row, by scoring the best finishing times in the three longest races. For years, Barb held the record for the most dominant performance in burro-race history, winning the Leadville twenty-one-miler in 2010 by such a huge margin that the guy in second place finished a full half-hour behind her. That was also Barb's first race in two years after coming out of retirement—at age fifty-four.

But being best will beat you up. By 2014, Barb was thinking it was time to re-retire. The downside of running with fireballs like Chugs and her newest burro, Dakota, is you often don't have a say in how hard you hammer the descents. Barb always had a blast cannonballing down those long, rocky drops, especially when it meant blowing past Hal Walter and Tom Sobal, but sooner or later, twenty years of recklessness leads to a reckoning. Barb's hips and knees were feeling trashed, and all those big winter miles in below-zero windchill were getting to be a grind. Barb and her husband had a sweet little farm on the outskirts of Leadville, just a short mountain bike ride from Twin Lakes, where Barb could finally ease back and spend her time pedaling the foothills and hiking Hope Pass.

One thing bugged her: what about Chugs and Dakota? Chugs still had plenty of steam left in him, and Dakota was just hitting his stride. Now that she'd created such fine-tuned racing machines, Barb hated to leave them just standing around in her backyard. She wasn't sure what to do—until the dentist's office door swung open, and in walked Kelly Doke.

This could be a *reeeeallly* bad idea, Barb thought as she led Lynzi into the corral.

This kid was so quiet, and crazy young, and skinny; Barb was pretty sure Chugs had packed away more for breakfast that morning than this gal had on her bones. Barb didn't want to scare Lynzi away, but seriously, it takes some muscle to handle a 700-pound runaway. Most burro racers start after they've reached peak strength, in their twenties and thirties, not before they've gone to their first freshman mixer. And what was the deal with Lynzi's health—was Barb going to be calling for help when this kid collapsed at 12,000 feet?

"Lynzi is tougher than she looks," Kelly assured her. "That's the only reason she's still alive."

Barb got it. All Lynzi wanted was a chance. So Barb handed her a halter and began sharing the wisdom that greats like Mary Walter and Diane Markis had passed down to her. Barb showed Lynzi where to position herself *(Get behind him and tuck in close, like you're joined at the hip)*, and schooled her in the Growl. "Lynzi, you can't just whisper 'C'mon, Dakota,' like you're feeding a kitten," Barb scolded. "It's got to come from your gut. *HhhheeeeyYAAAAA! YA!* Git goin'!"

Barb even unveiled her secret weapon: the Trash Wrapper Trick. The Growl will get a burro going, Barb explained, but it's wicked hard to slow them down. So early on in the Chugs years, back when Barb was coming home bruised and bloody after being dragged through juniper thickets, she began experimenting with an old technique from nineteenth-century Russia. During her long training runs, Barb told Lynzi, she would always carry a few Clif Bars in her pockets. Every time she stopped for a snack, she made a big deal out of opening the wrapper, crinkling and tearing it loudly, before sharing a bite of the bar with Chugs. Over time, the Pavlovian conditioning locked in: whenever Chugs heard the crinkle of a wrapper, he hit the brakes. If they were bombing downhill and Barb couldn't match Chugs's speed, all she had to do was reach into her pocket.

That spring and summer, Barb and Lynzi got together nearly every weekend. Sometimes Kelly joined them on her mountain bike; other times, Barb would have a little All-Star team reunion and invite along her best bud/archrival/multiple Triple Crown champion Karen Thorpe. No big deal, everybody, just a ninth-grade novice out smacking balls with Venus and Serena. Barb was always on Lynzi to speak up and take command of Dakota, but privately, she loved the girl's quiet strength. Talkers aren't listeners, and Barb had seen too many first-timers who acted like ugly Americans abroad, thinking all they had to do was shout louder instead of learning the language.

"You'll see people screaming, all this high-energy stuff, but the more you do that, the more you confuse a burro," Barb told Lynzi. "All you need is a few quiet commands. Burros give you a lot of signals, so you have to be in tune with that. Hal has that. Curtis has that. Burros have such big hearts, and that's what makes this sport so special. It's all about bonding."

Once, two of the world's best ultrarunners came to Fairplay to try their hand at burro racing. After triumphing in the toughest contests on the planet, it was time to tick this box on their bucket list. Max King had won the World Mountain Running title and the 100K World Championship, while his buddy Ryan Sandes had conquered basically everything else: the Leadville 100, Western States, all the major desert races, even the Grand Himalayan Trail.

Max and Ryan were paired with fast donkeys and tutored by none other than Meredith Hodges, the Trainer of Trainers, a mule and donkey specialist with such command of the equine mind that even Hal Walter seeks out her advice.

"If you get too cocky, they will most definitely humiliate you, and they'll pick the biggest crowd to do it," Meredith warned Max and Ryan. She taught them the rudiments of burro communication, then turned them loose to try their luck in the World Championship Pack Burro Race. "If these guys have really done their homework with their donkeys, if they've really fostered that relationship in a positive manner, they will be running together," Meredith predicted. "If you engage in a partnership with them—if you are polite, considerate, respectful in the things you ask—and you keep it fun and exciting, they'll like doing things with you. They like the extra excitement. They think it's pretty cool."

On paper, it was the perfect marriage of animal and human. On race day, it was a disaster. Ryan and Max spent a good part of the morning pulling their donkeys up the mountain, trudging like galley slaves with ropes over their shoulders. Ryan finished in sixth place, well behind weekend warriors like Caitlin Jones, and Laura Hronik, and fifty-seven-year-old, AARP-eligible Hal Walter. Max was even farther back in sixteenth—out of sixteen. The pros could run like the wind, but they couldn't partner for crap.

Three months after Barb met Lynzi, she cut her loose.

For her last lesson, Barb taught Lynzi how to sling a packsaddle over Dakota's back and cinch it just right (loose enough for comfort, secure enough to sit tight) and how to tie down the required pick, shovel, and mining pan. Then Barb sent her off to the first race of the 2014 season: an eight-mile tune-up event just west of Denver in Georgetown. The gun blasted, and suddenly Lynzi was swallowed by the stampede. She emerged ninety minutes later, finishing in the middle of the pack but with such a big smile and loping ease that she clearly had plenty left in the tank.

Lynzi fights the men for the lead in her first burro race, and lands on the front page.

"People began to notice," Barb says. "They're asking, 'Who is this sweet young gal?' "

They soon found out. Right before the Fairplay race, Barb made a strategic switch. She knew Lynzi had never run fifteen miles in her life, let alone strapped to a donkey, so Barb decided to pair her with Chugs instead of young, fast, and feisty Dakota. Dakota was speedy enough to give Lynzi the run of a lifetime, but Chugs had mellowed over the years from a half-wild thunderbolt into a wise, reliable warhorse. Barb couldn't stomach the thought of something going haywire with Dakota while Lynzi was up on the mountains: alone, exhausted, maybe hurt. Barb had been there herself: once during a race Dakota had gotten wedged up to his chest in a snowdrift, and Barb had barely been able to dig him out. She hoped Chugs was canny enough to keep Lynzi out of trouble.

Lynzi felt her stomach heaving as she and Chugs squeezed into the braying, stamping donkey herd. She tightened her fist around Chugs's rope and tried to relax. Shortly before the final countdown, she caught the attention of America's least-likely burro racers: Rick and Roger Pedretti, two middle-aged Wisconsin cattle farmers who haul their burros nearly 3,000 miles round-trip every year to compete in Fairplay. Fifteen years earlier, a Pedretti actually won the World Championship. Rob Pedretti had left the family farm to work in Colorado as a hunting guide. He didn't know it, but he'd been waiting for burro racing all his life. His three great loves were animals, mountains, and running, and when he discovered some genius had actually combined them in a sport, he was ecstatic. It took him a while, but by 1999, Rob even got good enough to defeat defending champion Hal Walter and win the World Championship. Five years later, Rob returned home to Wisconsin, stomped "I LOVE YOU ALL" in giant letters in the snow, and shot himself in the heart.

Every July since then, the Pedrettis have rented all the rooms in the ancient Hand Hotel for race weekend and made the World Championship their family reunion. The Pedrettis weren't runners before—far from it—but they sure as hell are now. The brothers all take turns running with Rob's burro, Smokey, always accompanied by a pack of Pedrettis of all ages with various burros borrowed from Rob's old friends. The race has become such a rite of passage, even hopeful boyfriends know the best way to impress the family is to grab a rope on race day and pay their respects to Uncle Rob.

So when Roger and Rick Pedretti spotted a scared teenager just before the starting gun in 2014, they knew exactly what Rob would have done. "Put Chugs between our burros and we'll stick with you," they told Lynzi. Or at least they'd try. When the three of them rounded the halfway turn and began heading for home, Lynzi looked so fresh that the Pedrettis suspected they were holding her back. "Don't be polite," Roger said. "If you can go, get going!" Lynzi was off like a shot, picking off racer after racer as she and Chugs weaved through the pack. She sprinted across the finish line, the number one woman and the fourth overall in her very first World Championship.

"That gal is the real deal," an impressed Curtis Imrie remarked.

"Latest in a long line of women in this sport who are incredibly fit, very competitive, and have the knack for training animals. Why wouldn't they beat all the men?" Curtis had a saying he liked to spring on new burro racers: "You have a great past ahead of you." But with Lynzi, he saw a whole new future. "She has the opportunity to redefine this sport," Curtis said. "Just the way Barb Dolan did."

Maybe she did. Or maybe Barb wasn't done defining it yet herself.

Because a few weeks later, spectators crowding the streets of Buena Vista heard a trailer door clang open and saw two legends emerge: Barb Dolan and Smokey, Rob Pedretti's champion burro. At twelve miles, the Buena Vista course is comparatively short (Curtis calls it "the track meet"), so Barb figured it was safe to team Lynzi with Dakota. For herself, Smokey was perfect: a pursuit demon who's obsessed with catching every burro in front of him. At the gun, Barb and Lynzi blasted off together, clattering across the bridge and leaning into the long climb up Whipple Trail.

They hit the peak side by side and whipped around the turnaround, Lynzi sucking deep for wind as Barb gave her a master class in How Shit's Done. Dead ahead was the front pack, led by a notoriously fast trail runner from California. The Californian was hooting and hollering like he was leading a cattle drive, knowing he was a lock for first and its $500 cash prize as long as the burro he'd rented kept pace. Barb glared at him, annoyed but minding her own business. Then she checked to see how Lynzi and Dakota were handling the commotion, and saw the future Curtis was talking about.

Lynzi was running the way we do in our dreams, as light and smooth as if the earth were spinning beneath her feet. She was tight at Dakota's hip, exactly where Barb had taught her, and encouraging him more by her presence than her voice. Every once in a while, Lynzi breathed a command. Dakota twitched his ears, listened, and obeyed. After watching Lynzi, Barb spun back to the Californian. "I don't mean to be a bitch," she said, "but shut up already. Keep it down, will ya?"

Her message to Lynzi was a lot shorter: "Go!"

Lynzi was gone. She dropped Barb on the long downhill, sticking hard on the heels of the lead pack as they thundered toward the finish line. Lynzi sprinted by past champions Bobby Lewis and George Zack, and was closing hard on Hal Walter and Justin Mock, the 2:29 marathoner, when she ran out of race course. Once again, the ninth-grader was first woman and fourth overall.

"How'd she do?" Barb asked, panting in a few minutes later. She smiled when she heard the results. "Pretty awesome. That's pretty awesome." There would be other days to run down the loud Californian and the rest of the guys, Barb knew. Lynzi's past was just beginning.

22

Skirt and a Smile

Coach Eric doesn't doink around. When I called him in a panic after Tanya's accident, I had a hunch it wouldn't take long for his big brain to come up with a battle plan. I still had no idea how we were going to get three donkeys from Pennsylvania to Colorado, but at least I had two more months to deal with that mess. Right now, our hot button problem was those three interlocked riddles:

- How do you prepare in a valley for a race in the mountains?
- How do you train alongside donkeys, yet stay one step ahead?
- How could three very different athletes like Mika, Zeke, and me learn to run together as a team?

Eric cracked the code in one night. "I've got a thirty-second solution for all your problems," he told me the next morning. "And you know what it is too."

I did? I was lost . . . and then it all came rushing back. "Oh, no," I groaned. "Not that again."

"Yup," Eric said. "You can't beat a classic."

The Thirty-Second Drill. It was the first thing Eric taught me when I met him ten years ago. At the time, we were in a city park in Denver and I was wrapping up my interview with him for *Men's Journal*. "Well, everything you've said about running is good for *some* people," I said, echoing the advice I'd been given over and over by the podiatrists and sports-medicine physicians who'd treated my various injuries. "But guys like me aren't built for it."

Eric exhaled slowly, a model of Zen forbearance. *Here we go again.* "Look, your running isn't hurting you," he explained. "It's the way you run. Let's try something." Eric had me kick off my shoes, and together we set off on a barefoot jog around the park. "Now here comes the magic," he said. "When we reach that tree, sprint to the next one."

"You mean, like—" I fumbled, oddly confused by those simple instructions until I realized that the last time anyone gave me that command, I was at high school basketball practice forty years ago. From then on, I'd probably cartwheeled more than I'd sprinted, and I can't cartwheel. Who sprints anymore? Anyone who's ever ripped a hamstring racing a nine-year-old niece knows there's only one sensible way to run, and that's to find your groove and stick to it. Sometimes we go a little faster, often a little slower, but mostly we cruise at whatever pace lets us finish three miles with a comfortable degree of discomfort. *Nobody* sprints. Talk about a recipe for disaster.

"Fast as you can," Eric insisted. "Let 'er rip for about thirty seconds. Then settle back to a jog."

Twenty seconds in, I flamed out. I had to stop and walk it off, wheezing like a guy being drowned in a bathtub. I hadn't sprinted in so long, I'd forgotten how. It was equal parts embarrassing and depressing, but Eric didn't give me time to mope. As soon as I recovered, we went at it again. And again. And by the fourth or fifth repetition, I noticed a weird sensation, like your hand tingling back to life after you've banged your elbow: instead of getting tired, my legs felt looser, stronger, *fresher*, than when we'd started. The faster I ran, the better I felt. I was straightening my back, driving with my knees, deep-breathing from the gut, and whipping my legs around from the hip.

"Good, right?" Eric asked. "Feeling bouncy?" Running fast can auto-correct your biomechanics, he explained, while slow leads to sloppy. That's a big reason I was always hurt; my plodding pace had me balancing too long on each leg, leaving all those tissues and tendons exposed to serious torque as my body weight swayed around. Instead, I should be *pop-pop-pop*ping my feet, getting them off the ground as quickly as I could.

"It doesn't mean you've got to sprint all the time," Eric said. "But the technique is the same. You've got to learn to go fast before you go slow." The Thirty-Second Drill was kind of genius: it was a workout, biomechanical feedback device, and fitness tracker all in one. And it couldn't be simpler: first, you warm up with an easy two-mile run. Then you sprint for thirty seconds, and jog lightly to recover. Repeat, alternating sprints and jogs, until you've had enough. Don't worry, you'll know when; as soon as your legs lose their bounce and you're struggling to recover between reps, that's the time to call it a day.

For the next nine months, Eric used cadence drills and speedwork to prepare me for the fifty-mile Copper Canyon ultramarathon, and it was a game-changer. But once that summer was over, so were my big training challenges. I stopped following a workout schedule, and gradually I regressed back to my slow and comfortable groove. For the next ten years, my speedometer barely quivered. Sprint? C'mon. Nobody sprints.

Eric's new battle plan, despite the sinister shadow of the Thirty-Second Drill, looked pretty good.

It was a Four/Two/One system: four days with the donkeys, two days without, one day of rest. Every time we ran with the donkeys, Eric wanted us heading one way: up. All hills, all the time. That way, we'd be building our wind and our leg strength while teaching the donkeys to ignore our gasping. Rather than pretend we weren't dying, we had to normalize it. Flower needed to get used to the idea that most times, the guy behind her would be blowing steam like the Little Engine That Shouldn't Get Its Hopes Too High. In Colorado, we'd be at 12,000 feet and fighting for air even when walking, so the donkeys had to learn that wasn't a signal to stop.

Non-donkey days were all about overdrive. Months ago, Eric had given me a workout schedule based on lots of short, fast repetitions, but once Zeke joined the team and we'd gotten serious about donkey training, I'd let that slip. Dumb mistake; we'd lost a good opportunity to stay one step ahead of the Gang of Three. But Eric believed we could make it up by adding a burst of speed whenever we ran on our own. The more we pushed the pace, the more we'd lower our resting heart rates and improve our performance at high altitude. It wouldn't completely acclimate us for the Rockies, but it would approximate the sensation and act as a good boot camp to prepare us for the stress of oxygen debt.

That sly dog Eric also knew something else about hills: They're Mother Nature's best remedy for big egos, second only to donkeys. Hills are the universal equalizer; that's why even in races, shrewd ultrarunners will hike any terrain that makes them lift their heads. Sure, a runner will beat a hiker to the peak, but not by much—and not for long. Three or four climbs into a 50K, a runner's legs will be trashed, while the hill hiker still has enough energy on tap to break away on the downhills and flats. Hills are more a test of shrewdness than stamina; you've got to have the experience to realize that your best climbing speed isn't much faster than anyone else's and the humility to accept it.

Eric knew that if Mika, Zeke, and I ran a lot of hills together, we'd find that all six of us—men and women, young and old, animal and human—had roughly the same sweet spot. We wouldn't have to worry about pacing together as a team: the hills would take care of that for us.

· · ·

Mika and Zeke liked Eric's game plan. The donkeys freaking loved it.

The day after Eric laid it out for me, we roped the Gang of Three and set off for our first hill workout. I wasn't looking forward to Phase 1 of persuading Flower she's supposed to keep climbing when I'm lagging behind, but as we were heading down the gravel road, I got an idea. I veered right, steering Flower toward the creek. She stopped short at the bank, letting Mika and Matilda catch up. "Let's take them across here," I suggested. "I want to try something on the other side."

Matilda had never been in this part of the creek before, but it didn't matter. She plunged straight down the bank, splashing through like a kid kicking puddles. Sherman was quick to follow, which meant this time it was Zeke who found himself dodging a 700-pound meteorite when Flower realized she was being left behind and leaped frantically to follow. The three donkeys scrambled to shore, poking their heads through a thin veil of weeds as they approached this trail they hadn't seen before. Flower swiveled her head like a radar scanner, ears set on maximum alert, as she sniffed for danger. Her nostrils issued a command:

Go nuts.

Flower upshifted from zero to holy-shit in about six steps. The rope whistled through my hands, stopping only when I bolted after her in time to grip the knot at the end. All three donkeys took off like a pack of wildebeests, galloping so crazily that I gave up any hope of dodging branches and just crashed along in the middle of the herd. Sherman and Matilda were galloping hot on Flower's tail, but I was so close to wiping out that I couldn't risk a glance back to see how Zeke and Mika were doing.

"Everyone good?" I hollered.

"YAH-*WHOOOOO*!" Zeke howled from somewhere nearby.

"Yoowooooah!" Mika hollered in the distance. Flower thundered on, zigzagging like a slalom skier along the snaking trail, forcing me to leap wildly over a gully I barely spotted in time. Only when Flower finally slowed down on a muddy uphill did I have a sec to wonder why, in twenty years, I'd never heard Mika wolf-howl before. Come to think of it, how could Mika be that far behind when Matilda's snout was right beside me? I looked back and saw Mika's rope but no Mika.

Crap. That wasn't a howl; it must have been a scream of pain. I dropped Flower's rope and sprinted back down the hill, only to find Mika hiking toward me. "Wait, I'll come to you," I called. "Is it your ankle?"

"No, I'm fine," she said. "I was saying, '*You go!*' "

Matilda had sprinted out so fast that Mika decided—smartly—to drop the rope rather than risk getting yanked off her feet. Matilda is only half Flower's size, but she can really turn on the burners. When she makes up her mind, you don't want to be in her way. "Oh my god, she was so excited," Mika said. "Sherman and Matilda were jumping and kicking like they were on the playground."

That was my plan, although about 50 percent less bananas. When we set off that morning, I knew the sticky part of Eric's strategy would be pushing the donkeys forward while we were lagging ass behind. Unfortunately, Flower was the key: she was pace leader, so for the operation to succeed, the big skittish baby needed a lightbulb to flash between those furry ears and show her that this was all an awesome game of Keep-Away and the hills were her time to shine. We had to make the climbs irresistible and fun, the way Nancy Sweigart* did when she'd goose Bubba the Goat so he'd chase her around the yard. Or like that time we showed Sherman the dirt trail to the horse farm and he suddenly took off like a thoroughbred . . .

Dirt. That was it. That was the thumb-in-the-goat's-butt we were looking for.

Somehow I'd missed the connection between the way the don-

* From Donkey Tao, chapter 9: Because nobody really has any idea how you're supposed to teach a goat to race, Nancy invented her own method. "I'd sneak up behind and goose him. He'd take off running and I'd chase him. Then he'd chase me. Then he'd jump on the car and dance around on it till my husband came out and we had to stop."

Lest you look down on butt-thumbing and carhood rave parties as crude and unscientific fitness strategies, consider that Nancy went on to reign as a three-time Grand Champion who never missed a race for fifteen consecutive years.

keys ran and where they were running. I'd been so busy bitching to Eric about how aggravating Flower was on the asphalt road leading to the maze, I'd forgotten to mention how giddy she was once we got inside. The secret ingredient we were looking for—the flash of inspiration that would override Flower's caution and transform her into a hill-running demon—was literally at our feet. I kept thinking we had to jack up our mileage by sticking to smooth, paved roads. But that morning, it finally clicked that when it came to hills, we were better off lost in the woods.

The rest of May, we lived in the maze.

Flower quickly picked up on our new routine. I didn't even have to steer her toward the creek anymore; within two days of trying out that new trail, she began beelining for it as soon as we'd gone half a mile down the gravel road. As much as she mistrusted water, she adored the playground on the other side. The feel of dirt beneath the donkeys' hooves must have put jumper cables on some dormant DNA from their ancestral past on the African savannah and shocked it back to life, because as soon as they climbed that creek bank, they went wild. We'd sprint along with them as best we could, sometimes keeping up, sometimes dropping the rope and letting them go, always knowing that eventually we'd find them waiting somewhere ahead on the trail.

We'd pop out of the woods at the foot of the Big One, the long climb up Slate Hill Road leading to the maze, and right away I'd feel the bull's-eye on my back. Flower and I led the way, with Sherman and Zeke champing at our heels. I could almost feel their breath on my neck, and I knew why: Sherman wanted to stay close to Flower, but Zeke was dying to shoot ahead. He and Sherman had become an amazing team over the past few months, mostly because Zeke had become a wizard at reading Shermie's signals. Sometimes on our days off, I'd find Zeke's car in our driveway but no Zeke. He'd be out in the pasture somewhere with Sherman, either showing him to friends or just hanging out on his own, sitting in the grass and sharing apple slices. Zeke's mom used to secretly tail him to make sure he was really going to therapy, but she breathed easily when he set off to see Sherman because she knew he couldn't wait to get over here. Zeke was getting to know Sherman better than any of us, but when it came to the Big One, Sherman had to figure out Zeke.

I knew what was going on, and I kind of dug it. We were flowing way better up the Big One these days, partly because of Eric's speedwork drills but also because Flower now knew that the quicker she got to the top, the sooner she'd be back in the woods. Sherman was doing terrific at hanging tough; he had to work twice as hard as Flower to stay by her side, but for him it was worth it because there was no place on earth he'd rather be. So why the *hell*, Sherm had to wonder, is this blond kid constantly busting my hump to get by her? Sherman couldn't see that Zeke wasn't gunning for him; he was gunning for me. I didn't have to look back to guess exactly what Zeke was thinking: "C'mon, man. I was a nationally ranked swimmer. I used to backstroke farther than this—and faster!—and now I'm stuck behind some fifty-four-year-old hogging the passing lane?" The Big One was the perfect place for Zeke to cut loose and show his stuff—except that his shaggy partner had different plans.

Tough luck, amigo. Welcome to burro racing.

Zeke always pretended he wasn't trying to pass us, while I pretended not to notice. I'm a Philly guy who cut his teeth playing pickup hoops, so shit talk is my mother tongue, but I never ripped Zeke about our secret daily showdowns, or even mentioned them. Something was going very right with him and Sherman, and I didn't want to ruin the spell. Every day that they both felt strong and spunky enough to come after me and Flower was a good day for all of us.

Luckily, the maze stays cool even when it's stinking hot, which was good news for us as May became June, and June became an outdoor sauna. We dodged the heat as best we could by sticking to the woods, exploring so many of the maze's twisting trails that after two weeks Flower knew the place better than I did. I'd hesitate whenever we hit a split in the trail, but Flower just surged on, never doubting which way to go and never making a mistake. Barb Dolan had told me that would happen ("If there's a rock or a tree stump that spooks them, believe me, they'll remember it a year later"), but I never really believed her till I saw it myself.

Our miles were keeping pace with the heat; as the thermometer climbed, so did our long runs, increasing in distance until we could comfortably handle ten miles. Or not so comfortably: one morning, we lingered a little too long over Zeke's second breakfast and our third coffee and didn't get out the door until nearly eleven. Zeke and I had learned our lesson from Mika and were carrying waist-strap water bottles, but it was such a steamer that by the middle of the workout, all three of us were dry.

"As soon as we hit direct sun everything spiraled out of control," Zeke said when we stopped for a break. "Like watching the *Hindenburg* in slow motion."

"Let's bust out of here and head for the waterfall as soon as Mika gets here," I said—except there was no Mika. A few moments later, Matilda trotted up alone. I left the donkeys with Zeke while I jogged back down the trail to make sure she was okay. A few hundred yards away, I spotted Mika with her head down and hands on her hips. "I think I'm done for today," she said. We took our time walking back to the donkeys, but between the heat and the lack of lunch, she was still feeling woozy.

"You guys go ahead and finish on your own," she said. "I'll walk it in with Matilda."

Before I could answer, Zeke piped up. "No way," he said, shaking his head nope, nope, nope. "We start together, we finish together."

The kindness was from his heart, but the words were from his Wednesday-night side hustle. Coach Eric had told us to add speedwork on our off days, and all three of us had found a way to embrace the speed but lop off the work. Zeke's approach was to follow the Bubba-the-Goat model of, essentially, berserkering around the backyard and dancing on cars. He'd become a member of a parkour gang* that trains every Wednesday after nightfall in downtown Lancaster. Zeke loved the way parkour *traceurs* use the city as their gym, training in alleys and climbing the walls of parking garages. He was still a novice, but already he was discovering that parkour moves like Thief Vaults, Muscle-Ups, Tic Tacs, and Double Kongs were adding a big boost of agility, stamina, and upper-body strength to his donkey running. Parkour prizes craft and camaraderie over competition, and as an awkward newcomer who sometimes lagged behind, Zeke was touched that seasoned *traceurs* always circled back to make sure he never finished a workout alone.

Zeke by night was now Spider-Man–ing around the city. Zeke by day, whenever he wasn't with us, was disappearing into the deep woods to squat in the creek. As his second side gig, Zeke had become a student of "the Iceman": Wim Hof, the Dutch cold-water guru. Zeke was intrigued by Wim's peculiar approach to depression, but what really hooked him was the science. If Wim's theories were legit, Zeke could use cold exposure in a way the Iceman himself had never envisioned.

Wim Hof spent most of his life as an unknown eccentric living on a houseboat in Amsterdam—until one day he was caught on camera leaping into a frozen canal to save the life of a man who'd crashed through the ice. The victim had to be rushed to emergency care, but balding, bearded Wim emerged relaxed and refreshed, as if he'd just enjoyed a dip in the pool. Which he essentially had; even in the middle of winter, Wim will cut a hole in the ice and slip in for a daily swim. He became so comfortable in deadly cold that

* You can see the parkour athletes of Lancaster in action in the piece I wrote for *Outside* magazine. https://www.outsideonline.com/1928031/concrete-jungle-worlds-best-gym.

he began tearing up world records: He swam nearly two hundred feet under sheer ice to set one mark; ran a marathon barefoot and nearly naked in subzero weather in Finland; climbed higher than Mount Everest's "death zone" barefooted and wearing nothing but shorts; and somehow *raised* his core body temperature while submerged in ice for more than ninety minutes.

During his lifetime of research and one-man experimentation, the Iceman came to believe that cold plunges were a lost secret of supreme health, and he makes a compelling case. Our ancestors were always chilled to the bone, he points out. Homes were damp and drafty, work was done out in the wind, and waterproof clothing was a madman's fantasy. To live was to shiver, but we did have one thing going for us: in a terrific bit of evolutionary jujitsu, our bodies adapted to the cold so that the same frost that could kill us could also make us calmer, stronger, and healthier. It all came down to oxygen: much the way you build a fire by blowing on it, you can crank up your internal furnace by sucking in air. That's why you gasp and shriek when someone pushes you into the pool; the cold shock triggers your lungs into hyperdrive, accelerating your circulatory system. Super-oxygenating your blood not only warms you up, but calms you down; and because a clear head helps you survive in dire straits, your brain quickly releases soothing hormones to help you (ahem) chill out.

Put it all together, Wim Hof proclaims, and you'll see why an occasional blast of cold should be treasured, not avoided. We inherited a tremendous gift, but we became so focused on constant coziness that we missed the connection between temporary discomfort and lifelong health. Instead of keeping our bodies adapted to cold, we cocoon ourselves in climate-controlled bubbles. We leave our toasty homes to travel in cars with heated seats to our jobs in thermostatically controlled offices, and if we exercise at all, it's in a gym that's so hot, even Child's Pose makes you sweat.

But big medicine is just a cold plunge away, Wim says. To prove it, he bared his arm and challenged doctors to make him puke. In 2010, Wim allowed Dutch medical researchers to inject him with a bacterial strain of E.coli that causes fever, vomiting, and headaches. Wim believed he could control his immune system with the same focused breathing that insulated him from subzero weather. He was injected . . . and felt dandy. So researchers at Holland's Radboud University Medical Center upped the ante and recruited twenty-four volunteers. Twelve were trained by Wim, twelve were not, and then all twenty-four were injected with the same noxious bacteria. Wim's students came out fine; on average, they had almost no reaction and showed higher levels of an anti-inflammatory protein, while the other twelve got sick.

The Iceman soon became a favorite space monkey for researchers around the world. He was fitted with a special ice-water suit and fed into an MRI to map his brain functions, then a PET scan to study his body tissue and capillaries. His brown and white body fat ratios were analyzed, his cortisol levels were graphed, his hormone secretions were measured. *Harvard Business Review* even zeroed in on Wim to find out why workers who take cold showers are less likely to call in sick. All of the studies dissected different parts of the Wim Hof puzzle, but their results essentially pointed to the same conclusion: if you want to burn fat, relieve depression, get stronger, increase mobility, and stay healthier, you might want to start with blue lips.

The secret, the researchers hypothesize, might be the breathing: Wim takes his students through a twenty-minute hyperventilation drill, which helps insulate them from the cold but may also trigger their sympathetic nervous systems and their immune response. For more than a decade, the Iceman's self-healing strategy has been embraced by people seeking help with obesity, diabetes, severe arthritis, and even the crippling cramps and tremors of Parkinson's, and the feedback has been so overwhelmingly enthusiastic that even top athletes, like surfing legends Kelly Slater and Laird Hamilton, have become disciples.

"Oxygen fires every cell in your body," Laird says. "Breath is what dictates the failure. If you look at any fighter, any athlete, as soon as they start mouth-breathing, panting, they're done. You can do without food for weeks, without water for days, but cut off the oxygen, and you're gone in minutes. Breathing is your power."

Zeke was trying to assemble the ingredients for his own Iceman hack when he hit pay dirt one afternoon in the woods. He was out for a solo run in the forest along the River Hills, and while following a creek that sliced beneath a soaring stone cliff, he found a deep plunge pool. He shucked off his running sandals and tested the temperature with his foot. The hole was so sheltered by trees and boulders that even on that hot spring day the water was as shocking as fresh-melt snow.

Oh my god, that's torture, Zeke thought. It's perfect.

Zeke slid in up to his chin. He stayed in the water as long as he could, then warmed up by running through the woods to the sunny Pinnacle overlook, the highest point in the Southern End. That evening, he went online and memorized the three specific steps of Wim Hof's breathing drills (thirty to forty power breaths; deep exhale and hold; deep inhale and hold; repeat for three more rounds). The next afternoon, Zeke was back in the woods. He lay on the creek bank while he executed his power breaths, then waded into the ice bath. *Brutal!* He wasn't sure if this would work for the long term. But short term, he'd certainly learned that your mind can't dwell on dark thoughts when it's screaming for you to get the hell out of this meat locker.

Zeke also knew that in London, an entire subculture of naked city swimmers had stumbled onto the Iceman's secret about three hundred years before the Iceman was born. In Hampstead Heath, London's giant public park, a cluster of spring-fed ponds was dug back in the 1700s, and ever since then, swimmers have trudged through snow to splash around in them even during the most bitter British winters. "After even a brief swim, I feel elated for hours and calm for days," explained Dr. Chris van Tulleken, a physician who was so invigorated by his own polar-bear plunges that he began experimenting with cold-water swims instead of pharmaceuticals as a treatment for some of his patients with depression. Sarah, for instance, was a twenty-four-year-old woman who'd been on antidepressants since she was seventeen but hated the feeling of living in a "chemical fog."

"She saw an immediate improvement in her mood after each swim and, as the weeks went by, her symptoms lessened," Dr. van Tulleken has said. Two years later, she's still off medication. Dr. van Tulleken shared his findings with two scientists who specialize in extreme environment performance, and together they published a study in *BMJ Case Reports* suggesting that a few frosty laps in the pool might be an effective treatment for major depressive disorder—which means those old Victorian codgers doffing their togs to break through the ice in the Hampstead ponds whenever they felt a bit of the "black bile" coming on might have had a better grasp of mental health than we do. Many of the outdoor swimmers interviewed by the researchers, in fact, said they began "in times of grief or bereavement and found comfort, even joy, in the water."

Zeke latched on to another tantalizing detail. The study's blockbuster news was the idea that a few minutes in a cold shower could be a remedy for such a deadly and baffling disease, but there was also a secondary effect. It was included in the case study almost as an afterthought, as if the researchers were apologizing because really, it had nothing to do with depression but appealed to them personally since they were experts on extreme performance. "Response to the stress of exercising at altitude is also diminished," Dr. van Tulleken noted. "This is called 'cross-adaptation,' where one form of stress adapts the body for another." Maybe, he speculated, learning to handle the shock of cold water would also "blunt your stress response to other daily stresses such as road rage, exams, or getting fired at work."

But that was thinking small. Zeke was thinking big. Cross-adaptation is more than medicine, he realized. It's rocket fuel. No wonder the Iceman was able to train dozens of amateur trekkers at a time and lead them on bare-chested expeditions up Mount Kilimanjaro at breakneck speed. Wim's success rate for these group climbs was astonishing: year after year, more than 90 percent of his students would reach the peak, despite the fact that many of them came to him because they'd been infirm or chronically ill. But here they are, stripping down to shorts and scampering nearly 20,000 feet up Africa's highest peak, and instead of suffering altitude sickness, they're hugging and high-fiving and breaking world records for group climbs. You never know what the mountain will throw at you, but by learning to breathe deeply, they'd found a way to conquer thin air, self-doubt, confusion, and exhaustion.

Wim didn't know it, Zeke told himself as he shivered in the creek, but he had also come up with a blueprint to build a better burro racer.

Maybe. But as far as Mika was concerned, the bottom of Zeke's swimming hole could be covered in pirate chests and platinum cards and there was still no way in hell she was going in. She would do just about anything to help Sherman, but after growing up on the beaches of Oahu, she drew the line at cold plunges in Pennsylvania creeks. Besides, she'd already found a guiding light of her own. Mika had become fascinated by Krissy Moehl, the champion ultrarunner who began beating men after she stopped thinking like them.

Krissy was twenty-two years old, fresh out of college and working in a running-shoe store in Seattle, when she began exploring the trails around town with one of her coworkers: Scott Jurek, the super-talent who would go on to win nearly every crown in the sport. Krissy didn't have big miles under her belt at the time; as a collegiate 800-meter runner, she was used to races being over in two minutes. But she had so much fun on those long rambles around Alpine Lakes in Olympic National Park that it wasn't long before Scott persuaded her to jump into her first 50K. Even before the starting gun, Krissy realized she'd made a big mistake. Track had taught her that racing is agonizing and unforgiving, a self-esteem killer that torments your nerves the day before and your body the day after. What kind of beating she'd get from five hours in the mountains, she didn't even want to imagine.

Krissy didn't need that misery anymore. For the first time in her life, she wasn't competing for her school, her coaches, or her team. She was running only for herself. So she should enjoy it, right? But was it possible to push hard and have fun at the same time? Krissy gave it a lot of thought, and decided the only chance she had of sticking with the sport was to follow three rules:

#1 *Smile from gun to tape.*

The more joy you put out, the more you'll see reflected back at you. Plus it was the best way to fend off Krissy's mom, who'd promised to make her stop if she didn't look alert and "with it."

#2 *Make someone else smile.*

When you're thinking about someone else, you forget how bad you feel.

#3 *Race like a demon.*

There's no fun in just plodding along, right? "Make no mistake, I count the ponytails in front of me," Krissy says. "But only after I take care of #1 and #2."

So before the running world knew Krissy's name, they knew her face. Spectators pointed as she flew into the aid stations, asking one another, "Who's the smiling girl?" She climbed the hills above Bellingham, looking down at the sea of mossy green firs sloping to the Pacific, and even though a vicious rain was blowing as she reached Mile 22 and had to climb a nasty hill called the Chinscraper, Krissy was elated. *This is my place*, she thought. *This is where I'm supposed to be.*

Krissy's rules were more like laws of physics: as long as she obeyed them, she was unstoppable. For the next few years, she ran the mountains like a thing unleashed. In the 2007 Hardrock 100, only two guys could beat her. At Hawaii's notorious HURT 100, only one. And *no one*—man or woman—could defeat her in a 100K in Oregon. But perhaps the greatest achievement of the Early Krissy Era was her assault on the Grand Slam: in eleven weeks, she ran all four iconic American 100-milers, becoming the youngest woman to complete the series and the second-fastest *ever.* (If you breezed through those numbers, take another look: in less than three months, Krissy ran *sixteen* mountain marathons.)

Of course, the happy-happy stuff couldn't go on forever. Krissy was becoming a bona fide phenomenon, and she began to realize she could actually make a living at the sport. Her big moment was coming up quick: the 2009 Western States 100, the premier American showcase for trail-running talent and the perfect opportunity for Krissy to become the first woman to win it all. She'd never felt stronger or savvier, and two fast guys agreed to pace her. Time to buckle down and get serious. On race day, she wiped the grin off her face, went out hard . . .

And hated it. She placed second among the women and thirteenth overall, but it wasn't her finish that disappointed her. It was the way she'd zombied through the whole day, feeling dead-eyed and anxious. She ran a hundred miles and enjoyed none of them, so focused on what she'd get at the end that she missed everything along the way. "I missed the beautiful sunrise," she realized afterward. "I wasn't smiling." She wanted to show her pacers she was just as tough as the guys, so she barely grunted as she charged through the aid stations and grabbed her drink bottles. She pushed harder than she ever had, reached deeper . . . and ran slower.

Krissy spent the next few days on her sofa, sore to the bone and wondering what she would do next with her life. Racing was over, obviously. If competing at the top of the sport meant putting yourself through that kind of torture, she decided, then no thanks. She replayed the race in her mind, starting from the final moments before the starting gun. She remembered huddling with the pack in the pre-dawn gloom, peering through the dark at the massive hill ahead: a four-mile heartbreaker shooting 3,000 feet straight into the sky. After that, you have ninety-six miles to go.

It really strips you down, Krissy thought. Running might be the only time in your life when you're not defined by how much you make, where you're from, what you did with your hair. "We're all stripped down to shorts and T-shirts," Krissy reflected. "Stripped to the core of who we are." Take her pal Scott Jurek, the most mild-mannered Clark Kent in the world. Ultrarunning has always been vexed by the mystery of what kind of weird voodoo spell comes over Scott on race day, when he tears off his glasses and suddenly becomes an absolute Superman savage, continually defeating the most ferocious racers in the world. At the starting line, Krissy recalled, Scott always leaps into the air and screams like Braveheart.

That isn't a show, Krissy realized. That's Scott. The core of who he is.

So who was she? Krissy stirred from the sofa, testing her aching legs. There was one way to find out.

Eight weeks after her crash and burn in California, Krissy flew to France for the biggest and most prestigious ultra in the world: the sadistic, soul-crushing, 106-mile Ultra Trail du Mont-Blanc. No American had ever won UTMB; few had even finished. Dean Karnazes, Scott Jurek, Geoff Roes, Zach Miller—all were at their peak when they attempted UTMB, and they all had their asses handed to them. Hal Koerner, one of the top American ultrarunners of the past decade, finished UTMB only once in three attempts, and even then it took him forty hours and a plastic bag around his balls because his testicles were chafed so raw. All ultra races are brutal in their own way, but UTMB is brutal in *every* way: you're frozen by night, scorched by day, tripped by scrabbly rocks, suffocated by nearly 33,000 feet of cloud-top climbs (imagine climbing Mount Katahdin—right after you'd climbed Mount Everest).

Krissy blew all that negative noise out of her mind. Six years earlier, she'd won a much shorter version of the UTMB, but for her attempt this time on the official course, she decided to reverse everything she'd done at Western States. Her mom was right; if you're not engaged and "with it," you might as well be somewhere else. So Krissy invited a gang of girlfriends to serve as her crew, and instead of hardcore performance shorts, she dug through her dresser until she found something cute—so cute, in fact, that during the race, Krissy could hear spectators in mountainside villages calling out, "That runner is wearing a skirt!" When she stumbled out of the woods and into the aid stations, she made sure to stop for a bowl of pasta with her gal pals and remind them—remind *them*—to make sure and get some rest. The sun rose, and set, and when it rose again, Krissy was still having a blast—and still in the lead. She snapped the tape to set a new course record, becoming the first American to win UTMB.

Krissy's skirt-and-a-smile style was so thrilling, so audacious and inspiring, that a pediatric nurse in Sacramento wondered if it would work for her too. Rory Bosio worked long shifts in a pediatric intensive-care unit, caring for infants who were just a razor's edge from death. The last thing she needed in her life was any more stress, so whenever she was off the clock, she couldn't wait to get outside and play in the woods. Rory Nordic-skied all winter and ran trails all summer, spending so much time on the rocky heights that her family called her "Billy Goat." When she tried her hand at ultrarunning, she was a natural—but just like Krissy, when she got to the big show, her soul hit the wall. Rory trained so hard for the 2010 Western States and raced it so fiercely that for months afterward, she could barely walk up stairs without gasping for air. Her doctor found she was so anemic, he recommended a blood transfusion.

Are you kidding me? Rory thought. She was stuck in Aesop's smuggest fable. She was allowed to *like* running but not love it, because if she loved it too much, she'd lose it altogether. The Billy Goat's short, glorious racing career was over before it had really begun.

Unless . . .

Maybe it was time for someone else to steer the ship. She'd tried letting "Rory the Disciplined Nurse" and "Billy Goat the Hard-Charging Hill Marauder" run the show, and those two had landed her in the hospital with a needle in her arm. All she had left were long shots, so she might as well throw the dice and hand things over to her *alter*-alter ego, the side of her that had gone underground since her days and nights of college partying: Welcome back, Bozo.

You think Bozo *trains*? Get real. Bozo plays. Whenever there was snow, Rory would strap a mini-sled to her back, run up the biggest hill near her home in Truckee, California, and bomb back down on her belly. On warm days, she'd churn across the lake on her stand-up paddleboard, or head to the park to rock out with her Hula-Hoop. Once a week, she had a regular date with Alejandro, her chunky but beloved beach cruiser, and together they'd crank all eighteen miles to the top of Donner Pass, one single-gear pedal push at a time. Along the way, she'd treat herself to snacks, reaching into her sports bra to pull out the Baggies of boiled sweet potatoes and avocados she'd stashed. Rory began to live the advice she'd gotten from a good friend: every day should be "a grand adventure where you're in the backcountry doing what you love."

Three years after Rory had been broken by Western States, Bozo edged into the mob of 2,500 runners from nearly a hundred countries awaiting the start of the 2013 UTMB. By the next morning, 2,493 of them were still suffering in the mountains while Rory was leaping madly across the finish line and then—to the roaring delight of the crowd—twirling to thank everyone with a ballerina curtsy. Her performance was mind-blowing: Rory was the first woman and seventh overall (!), and her time of 22 hours and 37 minutes shattered Krissy Moehl's record by more than two hours. By comparison, Rory was a *full half-day* faster than Hal Koerner's best showing.

Rory stuffed her trophy into her luggage and headed home, returning to her twenty-four-hour shifts at the hospital and picking up where she'd left off with Alejandro. One year later, she was back at UTMB, but this time she was the Bozo to beat. No more slipping anonymously through the pack; all eyes were pinned to the bull's-eye on her back. But not for long: Rory smoked the international field for the second time in a row, becoming the first woman to win back-to-back UTMB titles. Skirt-and-a-smile had triumphed again.

So why were American women like Krissy, and Rory, and 2007 champ Nikki Kimball able to crush UTMB, while American men were limping behind with sandwich bags on their nuts?

Probably for much the same reason that Zeke and I cratered on our first run through the maze to Tanya's house. We were stronger than Mika, and faster—and when things got rough, we fell harder. We were more afraid of looking soft than we were of falling short, so instead of following her lead and playing it shrewd, we insisted on brute-forcing our way into trouble. Krissy Moehl runs into the same thing all the time. During races, she'll catch up to slower guys on a narrow trail and try to pass. Race etiquette dictates they should move aside, but instead, they'll speed up, so worried about "getting chicked" that they'll match her, surge for surge, until—inevitably—the wheels come off and Krissy is free to blast past. If these guys were smart, they'd step aside immediately and draft from behind—but testosterone ain't smart.

Mika is. She got a copy of Krissy's training book, *Running Your First Ultra*, and it became her bible. Mika still considered herself a dancer, not a runner, and here was Krissy Moehl basically saying *Perfect! Just the right attitude.* Mika looked to Krissy for guidance on her solo workouts, and soon her mileage and foot speed were ticking higher. The stronger Mika got, the more confident she became, and that self-assurance traveled down the rope to Matilda. The bossy little donkey could tell her partner was now in charge; during our runs I'd see her eye roll back toward Mika, ready for orders.

By mid-June, it was still anyone's guess whether our weird chowder of cold creeks, Thief Vaults, be-your-own-bliss, and thirty-second sprints was actually going to work, but the time for guessing was running out. Race day was in little more than a month, and we still hadn't found a driver. Before I threw myself full-time into that search, I wanted to know if it was really worth

Zeke and Sherm take a breather in the maze. Note Zeke's homemade sandals.

it. So early one Saturday morning, we put ourselves to the test. We gathered the donkeys and headed off toward our old nemesis: the Big One.

Four weeks earlier, we'd tried to run the Big One and it beat us. This time we were looking for payback—with interest. The goal was to run up and down the Big One six times and cap that with a round-trip of the maze. If we pulled it off, we were looking at a solid sixteen-plus miles, with about half of them straight uphill. Luckily, the morning broke foggy, so we were already on our third trip up the Big One before the heat began to bear down. We notched another climb, and another, and almost before we knew it, we were looking at one another at the top of the hill.

"I can't believe it," Mika said. She was flushed and breathless, but more like a lottery winner than a shipwreck survivor. "Are we still doing the maze?"

"You up for it?" I asked.

"I . . ." Mika paused, mentally scanning her body for signs of the system failures she was sure had to be there, but she couldn't find any. "Sure!"

Zeke was down on a knee. "You good?" I asked, before realizing he was adjusting the leather thong on the homemade gladiator sandals he'd begun using for his runs. Zeke had gotten so into the spirit of natural movement and Early American–style woodsmanship that he'd learned to make his own trail-running footwear.

"Absolutely," Zeke said. "Let's get Shermie out of the sun."

We paid a price for our bravado, struggling to push our sore legs up the maze's roller-coaster climbs, but it was worth it. We finally popped out and headed for home, cruising down the Big One on our way to a long, cool dip under the waterfall. We tied the donkeys in the shade and plunged in, so roasting hot it felt like steam was sizzling off our bodies. Mika and I lounged in the swimming hole, while Zeke—being Zeke—detected a structural flaw in the rough pile of rocks serving as a dam and began restacking them in accordance with proper Euclidean principles.

Hunger and a desperate need for BELTTS* eventually drove us out of the creek. We walked the last half mile toward home alongside the donkeys, still not fully believing what we'd done that morning. We didn't feel good anymore. We felt unfreakingbelievable. Zoned in. Completely on our game.

Until something snapped.

* Bacon, egg, lettuce, tomato, and Tabasco sandwiches.

23

The Tao of Steve Rides Again

Does this feel right to you?" I asked a guy named Nate, holding out my left hand.

About a month earlier, I'd added my own side hustle to Coach Eric's need-for-speed plan. I'd heard that every Wednesday night, a group of guys played two hours of full-court hoops in my daughters' school gym. I hadn't picked up a ball in thirty years, not since I quit cold turkey when I almost turned down that reporting job in Portugal because I was so attached to my home courts in Philly. It was kind of a rock-bottom moment, a sickening awareness that you're making decisions for all the wrong reasons. Now I was fifty-four, and most of the guys in this Wednesday-night game hadn't even been born the last time I played. Before I jumped in, I checked with Coach Eric.

"Great idea," he said. "Lateral movement, explosive power, short bursts of speed, upper-body work—all of it exactly what you need. Just don't get hurt."

"Not a chance," I assured him. "I'm only there to run the court. My feet aren't leaving the ground."

. . .

"What do you think?" I asked Nate.

I'd been playing defense under the basket when another player fell back against me. I held him off with my hand, jamming one of my fingers. I gave it a hard pull, trying to pop the knuckle back out, then forgot about it—until I went for a rebound and a million shards of molten shrapnel shot up my arm. I came off the court, grimacing and gripping my wrist, and went over to Nate, who was taking a break on the sidelines, for his opinion.

"You know I'm not, uh, exactly a doctor," said Nate. He's actually a customer rep for a local dental office, but he proceeded to examine my hand anyway. He probed it gingerly with a finger. "I'm not positive, but I don't think anything in your body is ever supposed to click. You're clicking."

"Probably just a little, like, dislocation," I said, more to myself than to Nate. It was June 22, exactly forty-one days until the World Championship. We'd overcome so much, trained so hard, and developed such a bond with the donkeys, I couldn't believe I might have suddenly ruined it all. No way. It couldn't be broken.

That was what I also told the emergency room doctor several hours later, when she came into the examining room with my diagnosis. "It's feeling much better," I said, holding up my swollen, black-and-blue hand. "See? It's just bruised."

She stared at me. "You know I have your X-rays, right? You're not going to talk me out of it." She slipped the films onto the big lightscreen.

"It's broken," she announced.

"Badly? Or like, hairline?"

"Take a look."

She pointed to something on the screen that looked like an exploded firecracker. One of the bones in my left hand had shattered in the middle, leaving the two ends too splintered to knit on their own. I was going to need surgery, she explained. They should be able to reconstruct the bone with six screws and a metal plate, but there was a bright side: If I healed quickly and really applied myself to physical therapy, I should regain full strength and mobility right away.

"Probably no later than the end of the summer," she concluded.

Mika and Zeke weren't giving up. We didn't have much hope, but we did have a plan.

Two days later, I went under the knife. The following morning, I watched from the window with my hand in a thick wad of bandages while Mika and Zeke roped up Flower and Sherman. They were going to run two donkeys at a time, a different pair every day, so all of them would stay in shape while I was on the mend. Matilda stood at the gate, stunned. She couldn't believe she was being left behind, then she made sure we knew how she felt about it by galloping back and forth along the fence line, braying bloody murder.

My assignment, meanwhile, was to start working the phones in a one-handed search for a cross-country truck driver, a livestock trailer, and, oh yeah—a place to board and corral the donkeys while we were in Colorado. *If* we were in Colorado.

Maybe Hal and Curtis could help. Hal Walter and Curtis Imrie were always getting themselves into impossible predicaments and seat-of-their-pantsing back out again. Once, Curtis was hired to portray a cowboy in a car commercial, and as part of the deal he got temporary use of a new Pontiac Fiero. "Temporary" is in the eye of the beholder, and Curtis decided to keep beholding that Fiero until they pried the keys out of whichever hand wasn't busy giving them the finger. For years, Curtis managed to dodge GM repo teams and summons servers, shifting the car from hideout to hideout and turning friends like Hal into outlaw accomplices by stashing it under tarps behind their barns, until GM finally surrendered. I was sure that with that kind of can-do spirit, plus all their experience at long-distance donkey hauling, Hal and Curtis would have some ideas.

Better find someplace quiet, I thought, feeling a little self-conscious about talking to Hal while one of our donkeys was airing her grievances in the background. But Matilda, I realized, had suddenly gone silent. I glanced outside to check on her, and found Mika and Zeke back in the driveway. "Forget something?" I called.

Mika's head jerked up with a look I'd rarely seen but instantly recognized. She was fuming. Zeke looked just as pissed. Sherman and Flower, on the other hand, were delighted. Their manes were swaying as they trotted toward the fence, happy to be reunited with Matilda.

"We got nowhere," Mika said.

"They just played us," Zeke added.

By the time Mika and Zeke had covered the hundred or so yards to the dirt road, they were so frustrated and overheated they felt like they'd gone five miles. Mika said it was as bad as our first days with Sherman, or maybe worse; Flower had learned from Sherman how to spin in circles, and Tanya wasn't around to straighten things out. We'd done too good a job of surrounding Sherman with friends; the three donkeys had become such a tight team, there was no splitting them apart.

"They won't run unless they're together," Mika said. "Nothing works."

But after taking a night to recover, Mika and Zeke were back at it the next morning. Mika had spotted their mistake and remastered their plan: the problem wasn't the donkeys, it was the lineup. Flower was the best frontrunner and Sherman was a natural sidekick, but even donkeys can use an emotional-support animal. Matilda was the key; she couldn't stand being left alone, but she was independent enough to leave, so as long as she was a constant in all the pairings, the two-donkey system might have a chance. The other two would never go anywhere if Matilda was behind the gate yelling for them to come back.

Zeke liked the Always Matilda approach, but added his own twist: Matilda would stay chill if Flower was with her, so Zeke suggested he and Sherm start off on their own and have Mika and Matilda follow. On race day, he reasoned, all those other donkeys and so many miles of twisting trails made it almost certain that we'd get separated, so it made sense to prep Sherman in advance so he wouldn't panic. Zeke's math was sound, except he didn't account for one factor:

The return of the Wild Thing.

Zeke opened the gate, never realizing he was about to release a supervillain whose powers for destruction had only grown while it was entombed in ice. Sherman wasn't a recovering invalid anymore; over the past few months, all the companionship, wild runs in the woods, and fresh grass had built up his strength and confidence so much that there was no longer any sign of his old fear and fragility. Sherm was a new donkey; but deep inside, the Wild Thing's survival instinct still smoldered. Sherman's brain was as formidable as ever, and now in possession of a far more powerful body. Zeke didn't stand a chance.

Mika watched from the window, then ducked away to spare Zeke's feelings. The whole episode was so painful, she didn't want Zeke to know there were witnesses. Sherman feinted and corkscrewed, backtracked and buffaloed, anticipating all of Zeke's donkey-wrangling skills and shredding them to pieces. Even with Matilda munching peacefully in the pasture with Flower and keeping quiet, Sherman wasn't following Zeke anywhere. Mika waited for nearly half an hour, then came outside to put an end to the battle. Sherman and Zeke were still in front of the gate.

The donkeys had laid down the law: United we run, divided we stall.

Four days after my surgery, the hand therapist unwrapped the bandages and we got a look at the damage. My hand was swollen and stitched from wrist to knuckles like a deflating football. The fingers had frozen into a claw; they were so stiff that when I was asked to pick up some colored beads, I could feel the stitches stretching and the metal in my hand protesting.

"It's all up to you," the therapist warned. "You have to work with this hand every day, or it will never move again. I see people all the time who rested too long and their mobility was gone forever."

I opened my mouth to ask a question, then snapped it shut. If she never *told* me when I could start running, then I wouldn't know it was too soon, right? Half-assed and self-delusional as that reasoning was, it still ignored one thing she'd made super-abundantly crystal clear: If I jarred that plate loose, they'd have to re-open my hand, remove the hardware, and rebuild what was left of the bone all over again.

But just to see for myself, that afternoon I went on a trial run anyway. I wore my plastic splint and held my arm awkwardly, protecting it as best I could. "Feels fine," I thought, but that meant nothing. With Flower, my left hand was my anchor point; that's how I held the coiled rope, using the right to pull in and release any slack. When I suddenly had to clamp down, my left hand was the first responder. There was no way I could control a 700-pound donkey with just one hand.

But what if I didn't?

One thing that always stuck with me from the burro race I'd tried years ago in Leadville was the way the best racers looked like ballroom dancers, sticking so close to their donkeys that their bodies brushed together. Tom Sobal and Karen Thorpe and Barb Dolan seemed to have worked out some kind of mind meld so they didn't have to exert control and just communicated with the wave of a hand and a nudge of the hip. It got me thinking: Our donkeys wouldn't run unless they were in a pack, but maybe we could turn that to our advantage.

"Let's get Zeke over," I suggested to Mika. "I want to try something."

I expected her to refuse for my own good, but I miscalculated. Mika and I had never been through a year like the one we'd just had with Sherman. Our *Top Gun* call signs for each other were "Nervous" and "Reckless," Mika being the one who brought the girls inside when she heard thunder and made them wait it out in the basement, while I once put six-year-old Maya behind the wheel of the car with her younger sister and cousin in the backseat and showed her how to gun it while I pushed from behind after we got stuck in the rain on a strange dirt road I'd driven down to see if it led to the Susquehanna (Mika's first question when we got home: "Did you also show her the brake?" Answer: "Um . . ."). But after all those months of loading live donkeys into our minivan, and cleaning the "bean" out of Sherman's penis, and storming through the woods with no clear sense of how we'd get back out again, we'd learned more about each other than we'd ever expected. Beneath the surface, the two of us were tougher and more cautious than we'd ever let on. If Mika was concerned about something, I knew she had a good reason. And if I wanted to take a chance, she could count on me to know the limits.

She made the call, and in no time, Zeke was rolling up the driveway and hopping out of the car to help Mika gather the three donkeys. Instead of my usual twelve-foot rope, I chose one that was only two strides long: a short six-footer. "Hey YUP," I called, and Flower was revved and ready, heading straight for the road at an easy trot. Normally I'd roam a few feet behind, leaving enough distance to spot and head off any oncoming Flower goofiness. This time, I stuck to her side, glued so close that her tail whisked my leg as we ran. I was holding the rope with only my right hand, which meant if she chose to bolt, I couldn't stop her. All I could do was let her know that I was there and it was time to do our job.

As we approached her favorite trail, Flower veered toward our usual creek crossing. I wasn't ready to risk slippery rocks yet, so I pulled back gently, ready to quick-release if she resisted. But she took the correction like a champ, looping back without breaking stride. She was caught off guard again at the end of the gravel road, expecting to go right and head up the Big One, but when I slid around and nudged left with my hip, she read it perfectly.

It felt pleasantly weird to be back on flat road after we'd spent so much time on the trails. I'd forgotten what it was like to actually run *with* Flower; in the woods, it was more like mutually beneficial bedlam, all six of us doing our own thing while playing the same

Get so close you disappear:
Mika and Sherman practice the art of Donkey Tao.

game. But today, Flower and I were a pair of Ice Capaders. Our bodies were so close together that unconsciously I'd let my legs and lungs slip into her cadence; our feet were hitting the ground and our breath was huffing out in perfect drumbeat rhythm. I could tell in a blink what she was about to do next; I don't know if I was reading the terrain or sensing some twitch in her body, but before she moved to the side, I felt it coming and shifted out of the way.

Flower and I were so keyed in to each other, it was as if she knew what I wanted before I even asked. Almost like I wasn't even—

Holy shit. The memory hit me so hard, I caught a foot on a rock and nearly tripped. One of the great things about running, even with a donkey, is the way your mind travels to places it never goes at any other time. And that morning, it suddenly hit me for the first time in nearly twenty years how wrong I'd been about that whole Tao of Steve philosophy I'd been swearing by since the night I laid eyes on Mika. Every time I'm asked how we met, I tell that story: "The three parts of the formula are 'Be desire-less, be excellent, and *be gone*,'" I would explain. "And I believed it so much, I literally grabbed my coat and left, turning down a ride home from Mika in a freezing storm because the Tao says get out of sight before your act gets old."

It took Flower to show me I didn't know dick about Tao. In a flash, the pieces snapped together in a way I'd never seen before, as if my brain had decided that this dirt road across from Roy Beiler's goat farm was the ideal spot for the end of an M. Night Shyamalan movie: "Flower loves to run," I thought, "and so do I. When two of you want the same thing, your own desire doesn't matter anymore. *You're desire-less*!" We practiced together and got good—nearly excellent, really—but one thing was still missing: that morning, I discovered Flower needed a little more reassurance. All she wanted was a touch of my hand, or a snort from Matilda, to let her know that when she was out there in front, she wasn't alone. When I kept my distance, we lost that connection. But when Flower knew I was near, I could disappear.

I was gone. I was desire-less. And it was awesome.

Now that Donkey Tao had been updated for Flower and was back online, we were soon back to building solid miles. We stayed out of the woods and stuck to lonely roads, trying to protect my hand by avoiding any treacherous creeks or trails. We'd still have time to sharpen our water crossing and trailrunning footwork in the final days before the race, I figured. For now, the priority was to get strong and avoid another fracture.

We still had one major headache looming—until out of the blue, a bolt of lightning struck. Waving a scrap of paper.

The lightning bolt's face looked familiar, and then it came back to me in a rush: the woman in my driveway yelling from her car window for me to c'mon over was Tanya's friend Shelley, the one who'd raced over to notify us about Tanya's accident. I felt a twinge of guilt—it had been a while since I'd visited Tanya—and then my stomach flipped. Uh-oh. Shelley's news was so urgent, she wasn't even getting out of the car. This couldn't be good.

"Iiiiiii've got something for yoooouuuuu," Shelley singsonged. She was trying to open the car door with one hand while waving a paper in the air with the other, but she'd forgotten to unfasten her seat belt. Both hands disappeared, then the door swung open and she popped out.

"Look what I've got!" she crowed. "Do you still need a driver for Colorado?"

"Don't tell me you found somebody!" I'd asked all of our dairy-farming and pig-raising neighbors if they could haul our donkeys, but none had either the equipment for that kind of drive or the free time to wait around until after the race to bring us home. I kept kicking the problem down the road by promising myself that in a pinch I could always send a Hail Mary to Hal Walter and ask him to road-trip out with Harrison for a father-son donkey-hauling adventure, except I'd never actually asked Hal if he was interested in a donkey-hauling road-trip adventure.

"Okay. So do you want me to tell you, or not?" Shelley taunted. "Because if you say 'don't tell me'—"

"C'mon, what've you got there?"

She handed over a scrap torn from a horse show program. Someone had scrawled "Call Karin," along with a phone number. Nothing more.

"I was at a horse show in Virginia, and I don't know how I got to talking about it, but I began telling this woman about Tanya's accident and how it wasn't just her that was in a jam, but you too, and all of a sudden she goes, 'Yeah, I'll do it.'"

"Wow. You think she's legit?"

"Oh, yeah." Shelley nodded. "I see her at all the big shows. She brings in these beautiful horse carriages from Europe. She's a very legit person. Don't know anything about her driving, though."

The second Shelley left, I began punching the number into my phone. A woman answered with a strange accent, a weird mix that somehow sounded authentically Deep South and totally phony. The drawl was unmistakable, but her "der" instead of "the" was weirdly German. I tried to draw her out with a little chitchat, but after telling me she lived in "V'ginia," Karin got straight to business.

"I hear you need someone to haul your mules."

"Actually, they're donkeys—"

"All der same. Me and m'girlfriends can take care of that for you. Just tell me where you got 'em and where you want 'em."

"Okay. Um—how much do you charge?"

"I'll figure it out," she said, as if eighty hours of hard driving behind the wheel of a two-ton livestock hauler was something she did for fun. "Won't cost you much. We'll sleep in the trailer."

No. This reeked of a scam, except I couldn't figure out the angle. This stranger wasn't asking for money; the only thing we'd be turning over to her were three mismatched donkeys of questionable disposition. Maybe she wanted to steal them? Sherman would certainly make her regret that decision. Plus, why go to all the trouble? From what I'd seen, donkeys weren't hot-ticket items. I'd gotten three for free in one year, and I wasn't even trying.

"Do you have experience?" I asked, groping for questions that would make sense of all this.

"Horses, plenty. Donkeys, no," Karin replied. "Always wanted to see Colorado, though."

Silence. My move. I wracked my brain, trying to decide which mistake would be worse: trusting a stranger to get Sherman, Flower, and Matilda to the race? Or missing the race because I didn't trust a stranger? "Let me think about this, and I'll get back to you," I offered.

"We ain't got much time," Karin said. "So if you want me on those dates, you gotta let me know."

That jolted me awake. Jesus Christ. The race was in three weeks. What other option did I have?

"Yes. Absolutely," I agreed. "We're on."

Karin promised to map a route, round up her girlfriends, and be at my door the Sunday before showtime. If they drove straight through, they would arrive on Tuesday. Just enough time for the donkeys to rest and for us to semi-acclimate to high altitude.

"Great," I said. "I'll talk to you—"

Click. Karin, whoever she was, was done talking.

For the next two weeks, I ran down a mental checklist every time we went for a run. I knew I was missing something crucial, but I couldn't figure out what. It was driving me nuts, because with the countdown clock at fourteen days, I was running out of time to catch and correct anything I messed up.

So what was I forgetting? We had three able-bodied runners again; well, able enough. Amos's brother had just trimmed the donkeys' hooves, and we'd even had the vet stop by to blood-test them for travel. "I'm used to puffy donkeys, especially when they've been in the barn all winter. Yours are so sleek and trim," she marveled. "Really wonderful." To Zeke's relief, all three got clean bills of health. "Can you imagine Sherman with rabies?" he said. "That would be apocalyptic."

We now had a driver, a trailer, three new packsaddles, and as of that morning, a place to stay. Thanks to the never-ending wonders of Airbnb, I'd located lodging for both us and the donkeys. Less than ten miles from the racecourse was a three-bedroom "Earthship," which we decided, after poring over the photos, must be a home that was carved out of the hillside and crafted from naturally reclaimed materials. Good enough! Even closer to town, a sheriff's deputy was renting barn space and pasture grazing at her farm. So the entire time the Gang of Three was in Colorado, they would literally be under police protection.

Maybe it was time to take a breath and stop worrying. It had been a crazy year, with one big hit rolling in after the other. But every time, we'd bounced back a little stronger, a little sharper and more united, than before. Something could still go wrong, but it was out of our hands.

Or so I thought, until Zeke's mom called. Damn! *That's* what I'd forgotten. I was supposed to check with the family that reserved all the rooms at the Hand Hotel to see if they had any space for Zeke's family. It completely slipped my mind until Andrea's name flashed onto my phone screen. Total fumble. Every room in the Hand had to be spoken for by now.

"Hey, Andrea," I began. "I know why you're calling. I'm really sorry."

"Oh." Andrea sounded surprised. "So . . . you mean Zeke spoke to you already?"

"Zeke? No. You're calling about the hotel, right?"

"No. Not the hotel." Andrea paused. "Zeke was going to call, but he's still processing it."

"Processing?"

"Yes. Zeke broke his foot."

24

Ladies ex Machina

Mika and I hurried over to Andrea's house and found Zeke slumped in a recliner, his foot strapped in a walking cast and a pair of crutches by his side. The TV was on, but Zeke's mom and two sisters had given up trying to engage him in the show or in conversation. Zeke was sunk in his own brooding world.

"Nice work, numbnut," I said, then saw from his face that it wasn't helping. "So what happened?"

After our run the previous day, Zeke said, he'd worked a double shift as a lifeguard at the Southern End public pool. The sun hadn't completely set when he punched out, and after eight hours of sitting in a chair twirling a whistle, he was dying to shake out the stiffness. He spent the last hour of daylight practicing parkour in the playground, kong vaulting over the picnic tables and cranking out muscle-ups on the swing set. By the time it was dark, he felt fantastic—and hungry enough to eat his own arm. He jogged over to the car, debating whether to blow off keto for one night and have a double burger on a bun, and that was when he caught the curb weird with his foot. He downplayed his limp when he got home, insisting to his mom it was only a sprain.

"But when I got out of bed this morning, I almost fainted," Zeke said. Andrea dragged him to the doctor for an X-ray, which confirmed what she suspected: shattered fifth metatarsal. Zeke's foot was as bad as my hand, except you need only one hand to run with a donkey. "So," I began, not sure if it was humane to even ask the question. "There's not any chance that maybe, in a week or two—"

Zeke's head shot up, as intent on his mom's answer as I was.

"Maybe what? If he can run?" Andrea said. "Absolutely not. That boot stays on for a month. *Minimum*," she added, staring Zeke down to make sure he got it.

"We'll see," Zeke muttered.

Mika and I were quiet most of the drive home. We were mentally orbiting the obvious, searching for something to say while circling the ugly fact that we wanted to avoid: This was checkmate. We'd finally taken a hit there was no bouncing back from.

"There's got to be somebody," Mika finally ventured. "Don't you know anybody in Colorado?"

"*Anybody* can't run with Sherman," I said. "Nobody can. You think Sherman will listen to anyone except Zeke?"

Only as the words were coming out of my mouth did their truth really hit me. Until that moment, it never really sank in how inseparable Zeke and Sherman had become; not until I realized that no matter how wide I threw the net, I couldn't think of anyone—no friend, no relative, no other runner—who could fill Zeke's shoes. Sherman was Zeke's donkey now, and you can't matchmake that kind of chemistry. Every so often I'd feel a stab of guilt when I was running up front with Flower, while Zeke, who was really a lot faster, was stuck in the rear, trying to make Sherman quit dillydallying already and stop messing around with Matilda.

But then I'd remember a few things: It wasn't Flower, I'd recall, who masterminded the mass escape of all the other animals by shaking the chain off the pasture gate with her teeth. It wasn't Matilda who figured out how to steal cat food from the woodshed by head-butting the door until it popped open. And it was

With Zeke out of action, Maya subs in to help train Team Sherman.

only Sherman who would take your arm in his mouth and pull you away, gently but damn seriously, if you were petting another donkey more than you were petting him. Shermie had a mind of his own and a will that wouldn't quit, just like the kid who spent all spring battling depression and trying to become a better burro racer by sitting in a freezing creek. Those two were meant for each other, and they knew it.

"You're right," I said to Mika. "We've got to do this." A lot had been thrown at Sherman and Zeke that year, and the two of them had soldiered right through. Zeke had been plopped into a crazy challenge, and he'd handled it with extraordinary patience and compassion; he never complained, he never gave up, and he never—except for that time we were all hangry and Zeke tried to push

Sherman with both hands like a shopping cart—lost his cool. I could only hope that every time Zeke sat down with his therapist, she was half as caring and dedicated to him as Zeke was to that donkey.

Mika was right. We had to find someone to give Sherman his crack at that race. Zeke deserved it.

One week later, I was peering out the kitchen window, cutting my eyes back and forth from the setting sun to the empty road. The day before, Mika and Zeke had caught a flight for Colorado with my younger daughter, Sophie, and my niece, Sara. The four of them were going to set up base camp at the Earthship, while I was supposed to help Karin and her girlfriends haul the donkeys.

Supposed to, because as of dusk, there was no sign of any horse trailer. In fact, I hadn't heard much from Karin since we'd had a little arm-wrestle on the phone a few days ago. I'd decided that even if I didn't have a choice when it came to drivers, I didn't have to hand over the Gang of Three to a complete stranger with no one to watch out for them. When I told Karin I wanted to come along for the drive, she sounded offended at first, then downright sour.

"Might not have room. Might have to put you back in the trailer," she grumbled.

"Fine by me," I shot back.

"Better be," she retorted.

After that—silence. The last thing I wanted was to annoy her so much that she'd quit, so I left her in peace and waited for her to show up Sunday morning . . . Sunday afternoon . . . Sunday evening—

BLAAAPPPP-Blap-Blap-Blap! A horn was screaming like I had five minutes to evacuate. Outside, someone was blasting away behind the wheel of a giant diesel pickup while, at the same time, maneuvering the biggest, longest horse trailer I'd ever seen between two trees and up my driveway. The driver finally took her hand off the horn. The front doors flew open, and down slid two women in culottes and flip-flops. I looked behind them, holding my breath in hopes that someone bigger, stronger, and younger was about to emerge from the backseat. Both of the women were as short as my eleven-year-old daughter, and if I had to guess the age of the one in the pink tee, I'd say sixty-something.

Nope. "Linda is seventy-two. Can you believe it?" said Karin, whom I identified by the odd accent and the fact that apparently she wasn't Linda.

"So, is this everyone?" I asked, still hoping. "You mentioned someone else?"

"Nah, Katherine couldn't make it," Karin said. "So looks like we can use you after all. We got a late start, so let's get going. Grab your gear while we load the donkeys."

"You better let me handle that," I warned. "They can be really difficult."

"Ha!" Linda snorted, stabbing a finger at Karin. "They ain't met difficult yet. Go on, get your gear."

Whatever. They might as well get a taste of Sherman right away and see what they're up against. I ran inside, quickly turning off lights and scribbling a note for Abe, the young neighbor who would be taking care of the cats, goats, sheep, chickens, ducks, and geese while we were away. I snatched my bag and locked the door, and in less than twenty minutes I came back outside to find Karin slamming shut the trailer door while Linda—seventy-two-year-old, eleven-year-old-girl-size Linda—was slinging fifty-pound bales of hay into the back of the pickup. I looked around for the Gang of Three. They were nowhere in sight.

"Wait. Are they in there already?" I said.

Karin waved me over to the back of the trailer. "Take a look." Inside, the three donkeys were all nestled in their own individual stalls, each separated by a half-barrier low enough for the donkeys to still nuzzle one another. Flower and Matilda were facing their windows, quietly munching from bags of fresh hay. Sherman, being Sherman, had his butt to the window and was staring stubbornly at the back wall. "He's your nervous one, huh?" Karin said. "Don't worry, we'll win him over."

I went to help Linda with the hay, but she was already finished. In less than thirty minutes, the two women that I'd been afraid couldn't handle the job had filled the water tanks, hoisted a half ton of hay, caught and loaded three suspicious donkeys, and I believe peed behind my car. Karin and Linda climbed into the front of the truck and I slid into the back, and before I even knew the plan—or if there was one—the two Virginia ladies were blasting us through Lancaster on our way to Colorado.

Sometime around midnight, Linda and I were filling the gas tank in West Virginia while Karin went inside for a rest stop and a Monster drink. "Something you should know about Karin," Linda said to me. "She won't tell you this herself. She had a really tough run with cancer a few years ago. Not up here"—she waved her hand across her chest—"but down here," and she pointed to her crotch. Twice, Karin battled uterine cancer, and the removal of a tumor "the size of a baby's head," as Linda put it, nearly killed her. Karin was left unable to have children, which also nearly ended her marriage. During that darkness, she vowed that if she ever got her life back, she was going to live every second of it.

"Now, nothing stops her," Linda said. "She gets an idea, she's gone. And me and Katherine are usually with her. If she calls us for an adventure, we grab our purses and go. Once, Karin found a jet boat for sale in New Jersey. You ever been on a jet boat? They're crazy! We took my husband's trailer and were halfway there before I remembered to tell him I was leaving." That was pretty much the only reason the Virginia ladies signed on for this trip: when Karin heard she could visit Colorado for the first time *and* watch a bunch of weirdos run around with donkeys, it was jet-boat time all over again.

Linda had survived some dark tunnels of her own. "My mother had four kids, one every June, and it was too much for her," she said. "She was only forty-three when she took her own life. Can you imagine how you must feel to be that young and believe there's nothing worth living for?" Linda was twenty-one at the time, and she never got over the aching sense of loss. She self-medicated by staying on the move, working as a long-haul driver for a trailer company and crisscrossing the Midwest in nearly nonstop seventy-two-hour deliveries. When she married and began raising her own family, she found an ingenious way to microdose her wanderlust: she became a horse midwife, which meant racing out the door in the middle of the night whenever she got an emergency call about a mare struggling with a difficult birth.

"Sissy Spacek, Ted Turner, Jane Fonda—I've helped all kinds of folks with their foals," Linda said. That's how Linda and Karin met, as Virginia horse lovers who ran into each other on the same backwoods trails. It wasn't long before they were taking long, rambling rides together along the James River—and, every once in a while, obeying the call to hit the gas and head for the hills.

Karin strolled out of the gas station, jangling her keys in the air like a game show prize. "You're up, cowboy. Ready to take the wheel?"

Hard gulp. "This is a lot of trailer for me," I said out loud, while my brain was screaming, *Are you nuts? We're on switchback roads at two in the morning in the mountains of West Virginia. We've got headlights stabbing our eyes and live animals in the back. This is where I do my trial run?* "I've never handled anything this big. I don't want to make a mistake that could get the donkeys hurt."

"You make a mistake, *you* get hurt!" Karin said. "You know how much this thing costs?" She was still holding the keys in front of me. "Now c'mon, you'll be fine. We got a picnic in the back, we got good music. You'll be fine."

I climbed reluctantly behind the wheel, while Linda cozied into a blanket in the back to catch some sleep. Karin dug into the cooler for our midnight snack. She and Linda might be as tough as any trucker, but they don't eat like one. I'd thought we'd be fast-fooding our way across the country, but they'd packed an amazing feast instead: grilled chicken and avocado on ciabatta, sliced bell peppers and apples, hard-boiled eggs, little bags of salad. Karin double-checked the navigator to make sure we were on course for Indiana in case she fell asleep, then settled back to chat.

"Did you know I'm Dutch?" she began, finally revealing the mystery of her accent. Her family still lives in Holland, where she was raised as a country girl who spent every second outdoors. Even as a kid, she liked working with her hands, and she developed such a flair for mechanics that she graduated with a degree in electrical engineering and became a technician for state-of-the-art copiers in the United States. "I was over here working on machines in the Pentagon and the National Institutes of Health," she told me. Her Virginia home was a manageable commute from downtown D.C., but Karin loved the farm so much that every hour she spent in the city felt like an hour lost. Karin's husband, Butch, is a farrier, and that put Karin in contact with lots of weekend riders who owned more horses than they could handle. Karin began hiring out as a trainer, and her firm but gentle touch and keen eye for body language earned her a reputation as a wizard who could unlock the potential of even the most troublesome animals.

"That explains how you got the Wild Thing into the trailer," I said.

"Which one's that? Sherman?" she asked. "Oh, he's a sweet baby."

Sweet baby? I couldn't get over the difference between Karin-on-the-road and Karin-spitting-nails-at-me-through-the-phone, and I began to suspect it was partly my fault. When she suddenly appeared out of the blue and offered to save my bacon, she must have naturally assumed I'd be grateful. Instead, I fumbled around with doubting questions, which to her must have come across as some mouthy man challenging her expertise and independence. Once she saw that I was happy to follow her lead, she was amazingly supportive, trusting me to handle her beloved trailer and even letting her guard down when I asked about her illnesses. She coached me through the first hour of driving, until she finally gave me the ultimate attaboy: "You don't need me in your ear," she said, then balled her coat under her head and went to sleep.

. . .

Daybreak caught us just outside Terre Haute, Indiana. I eased the rig off the highway and into a Cracker Barrel parking lot. Before we went inside to eat, or even visit the bathroom, Karin made sure we attended to the donkeys. We pulled open the trailer door, and there was Sherman—still turned backward, facing the wall and ignoring his hay and open window.

"That guy is such a hardhead," I said.

"Nah, you can't think that way," Karin said. "Animals don't do things out of spite. They're not trying to teach you a lesson. That's the biggest mistake people make with animals, getting this idea that what they do has something to do with you. You gotta get yourself out of the picture, and then you'll understand what's really going on."

Karin asked me about Sherman while we were cleaning out the trailer, raking out the old pine chips and spreading fresh ones under the donkeys' feet. I told her how we'd gotten the Wild Thing, springing him from the hoarder's stall and bringing him back to our place, where he remained dead-eyed and withdrawn until that goofy goat Lawrence befriended him. "See, doesn't that tell you what's happening here?" Karin asked. Flower and Matilda had lived with Tanya on Christmas Wish Farm, where they always had plenty of friends and time to romp outdoors. Sherman had been locked away on his own like a prisoner in solitary. He had to come up with a mechanism to cope, and that was what we were seeing: in confined spaces, Sherman had taught himself to zone out until it was over.

"He's not stubborn," Karin concluded. "He's scared. So we're going to keep checking on him, and petting him, and letting him know he's not abandoned."

Linda began mixing a tub of electrolyte mash for the donkeys, while Karin and I carried buckets over to the back door and asked a dishwasher if we could borrow a hose. He disappeared and returned with the manager, who personally connected the hose and dragged it out to us with such brisk friendliness that I had to wonder if walk-up donkey service in the Cracker Barrel parking lot during the height of the breakfast rush was just a normal part of her daily routine. "Hon, anything else you need, just ask for Marilyn," she told Karin, then hurried off to check on her packed tables.

"Isn't that something?" Karin said. "Animals bring out the best in everybody."

When we finally sat down to eat, it was the first time we were all awake together and not focused on the road—and that's when I discovered what I was in for. Linda gave me the first hint when our food came. When I reached for a fork, she grabbed my hand. "That can wait," she said. She and Karin joined hands and bowed their heads. "Thanks, but this isn't something I believe in," I began, but the Ladies kept their eyes closed, waiting me out. *All right, be polite*, I told myself. *They are your hosts. And we are in a Cracker Barrel.*

After that opening shot, it was game on. Fate had conspired to take the reddest and bluest of America and lock it in a steel capsule for thirty straight hours. We drove out of Indiana and on toward Kansas, talking about everything and agreeing on absolutely nothing. We faced off on Confederate flags, school prayer, and Black Lives Matter, and when I argued in favor of stricter gun laws, Karin told me about the pistol strapped to her leg. The closest we came to common ground was after we heard on the radio that Trump had bad-mouthed the parents of a soldier who'd been killed in action. "Republicans screwed up," Karin agreed. "They could've had Sarah Palin."

We skirmished across the prairies, at the same time sharing the Ladies' sandwiches and making sure the blankets hadn't slipped off whoever was sleeping that shift in the backseat. It was weird, the way the three of us saw things so differently yet liked one another so much. Karin and Linda charmed their way across the Midwest, finding new Marilyns everywhere we stopped and inviting kids to come pet the donkeys' noses. Karin mocked my taste for Doritos, yet made sure to pick up a fresh bag whenever we gassed up. When Linda saw my reading glasses, she demanded I hand them over. "Looks like you dipped them in bacon grease," she said, cleaning them thoroughly with a tissue. As the sun was going down over

Kansas, Linda wondered why we hadn't spotted a single cow anywhere in the state. "It's the Rapture!" she declared, cracking herself up. "The Lord took the most innocent amongst us."

It was just past our second midnight when we crossed into Colorado, and that's when the tone inside the truck got serious. Snowcapped mountains reared up in front of us, smack on our path to Fairplay, and thick fog was rolling down the slopes and blanketing the road. "Fun's over," Karin said. The asphalt was crunchy with frost, and visibility was so poor that we couldn't see approaching vehicles until their headlights suddenly blasted out of the dark and into our eyes. We climbed the mountains at a crawl, creeping through the icy gloom at forty miles per hour . . . thirty . . . twenty. . . .

At four in the morning, our cell phones lost reception and the navigator couldn't locate the Earthship's address. "Maybe that's it?" I asked, pointing to a dirt road in the middle of a lonely mesa. "Better be," Karin said. "Once we go in, there's no turning this thing around." One flickering bar appeared on my phone, so I quickly left Mika a Hail Mary message, telling her we were wandering in the dark and this was my only chance to call, so if she happened to wake up, please come out and flash a light.

We rumbled along, spellbound by the star show stretching to the horizon but wishing for any sign of human life, anywhere. Suddenly Karin pulled over. "I got a hunch," she said, which didn't sound good until she told me to swap seats and get behind the wheel. "If I'm right," she explained, "you can show your lady your stuff." I had a pretty good hunch of my own—that this sorry excuse for a road was only leading us deeper and deeper into trouble, but I went along with Karin, took the driver's seat, and shifted the truck into low gear. Moments later, Mika burst into the headlights, waving her arms and running through the dark, welcoming us in from the road.

25

"Fear that thing. Do that thing."

If you bring forth what is within you, what you bring forth will save you. If you do not bring forth what is within you, what you do not bring forth will destroy you.

—*Gnostic Gospel of St. Thomas*

Any sign of the Ladies?" I asked when I crawled out of bed around ten that morning, still so groggy I could barely open both eyes.

"They left," Mika said. "They came in around eight, had coffee with Zeke, then took off in the truck."

"Jesus. Where to?"

"All they told me was, 'Wherever we end up is where we're headed.' "

The Ladies' stamina was absolutely freakish. After driving nonstop from sundown Sunday until nearly sunup Tuesday, we'd then worked for an hour in the freezing dark, making sure the donkeys were okay after the long trip and then turning them out in

Made it! Sherman wakes up to his first morning in Colorado.

the meadow with plenty of hay, fresh water, and electrolyte mash. After that, the Ladies had to clear away and stow their gear so they could get to the king-sized mattress in the sleeping loft in the front of the trailer. Three hours later, they were back on their feet and antsy for adventure.

"How about the donkeys? Anyone check on them yet?" I asked.

"Yup, Zeke is there now," Mika said. "He couldn't wait to see Sherman."

I looked around, taking in my first good view of planet Earthship. When we arrived last night, I was so fried that I barely noticed a thing before falling into bed. We were actually staying in the Earthship's satellite cabin, Mika told me. It was a beautiful straw-bale home, cozy and sun-drenched, crafted from bales of straw coated with a thick layer of adobe mud. Something about the thickness of the walls and the smoothness of the sanded adobe made the whole place feel indestructible and incredibly comfortable at the same time, like a cave remodeled by the *Queer Eye* quintet. Through the front window, I could see the nearby Earthship:

Sherm checking out his Colorado digs

a long, low building nestled against a gentle slope, so perfectly contoured against the man-made hill that Martha Stewart herself couldn't have designed a snazzier end-time bunker.

Mika and I decided to go see how the donkeys were settling in. As soon as we stepped outside and began heading down the long dirt drive, we spotted the Ladies approaching—

On horseback.

"Sleeping Beauty!" Linda hooted. "Princess finally got his butt out of bed."

"We been up so long, we had time to steal these beauties," Karin chimed in. "Hurry and open the trailer so we can hide 'em before the cops come."

The Ladies waited for me to answer, then realized from my slack

The Ladies: Karin (left) and Linda saved the day, then "stole" some horses.

jaw and completely befuddled stare that yeah, as far as I knew they were perfectly capable of recreational livestock rustling. "Nah, we borrowed them," Karin said. They'd set off on a drive that morning to get the lay of the land, and as usual, it wasn't long before they'd befriended another Marilyn. When they spotted two fine-looking horses standing idle in a meadow near a house, the Ladies knocked on the door and asked if they could take them out for a few hours. For whatever reason, the owner allowed two complete strangers to saddle up his horses and ride off down the lane.

"We just came back for our beer," Linda said.

"And my gold pan," Karin added. Before leaving Virginia, she'd packed a prospecting pan in case they spotted any promising creeks. Because who wouldn't? The Ladies slid off their horses, disappeared into the trailer, and popped back out a few moments later with Karin's pan and an insulated saddlebag full of beer and sandwiches. "If we don't come back," Linda said as she swung back into the saddle, "we were never here." The Ladies clinked their beers, spurred their horses, and were gone.

Naturally, the sight of two mounted outlaws cracking brews in their driveway at ten a.m. was enough to draw the attention of Kristin and Kip Otteson, owners and builders of the Earthship, who came out to see what was going on. I was glad to meet them, because it gave me an opportunity to ask a question that had been on my mind for some time: What the hell is an Earthship? Basically, Kip explained as we all walked together to the pasture, it's a passive-solar home that uses massive windows and natural insulation to be self-heating and -cooling.

"Even here?" It was still cold in July, so I could only imagine how bitter it must be in winter.

"Best home I've ever had," said Kip, and that meant something. Kip was originally a Southern California surfer who moved north—briefly, he thought—to attend college in Tacoma, Washington. There he met Kristin, and after graduation they set off together to teach school in a tiny village in Arctic Alaska that was reachable only by bush plane. "Two hundred people lived there, and one hundred of them were kids," Kristin said. "Everything revolved around the school. I loved it." Life in the tundra was so wild and woolly that whenever Kip took the cross-country team out for a run, he had to strap a .357 Magnum to his chest. "Year before we got there, a guy walking down the street with his girlfriend was jumped by a bear," Kip said. "She ran for help and was back in seventeen minutes. By then, he was already half-eaten."

Kip and Kristin arrived as outsiders but never had a chance to feel that way, because the tight-knit community immediately embraced them. They had such a wonderful time that after three years in Alaska, they had to wrench themselves away to set off for their next adventure in Thailand. In both Asia and the Arctic

Circle, they soon came to realize, there was a spirit of togetherness that seemed to be dying out in mainland America. "It's so rare these days for an American family to have four generations living close to each other," Kristin said. So when her brother-in-law proposed his crazy dream for a pioneer settlement, they jumped on it. Kristin's sister had married Jon Jandai, a Thai eco-visionary who taught green warriors how to build homes out of earth-friendly materials. During a trip to Colorado to visit Kristin's parents in Loveland, Jon Jandai got so excited when he found these forty desolate acres for sale, he nearly burst into flames.

"We can pen up elk, then kill them as needed by running them into a pit-fall!" he told Kip.

"Yeah," Kip replied. "You realize all that's completely illegal, right?"

But once they adjusted Jon Jondai's dial to twenty-first-century American fish and game laws, he inspired Kristin's entire clan to pitch in. They all bought the property together, and then, under Jon Jondai's guidance, a family community began to rise from this mesa in the middle of nowhere. "He's an incredible worker," Kip said. "He'd put a yoke over his shoulders and carry buckets from the pond for our adobe bricks. He pushed us to do stuff we never thought we could." Kip and Kristin and their two kids were still splitting their time between Thailand and Colorado, but now that they'd completed two of the homes, they were ready to settle in full-time and start work on the third house, for Kristin's brother.

"Now I see what we really need," Kip said. "Donkeys. They look so cool out there." We'd reached the pasture, where Zeke was sprawled on the grass in his boot, soaking up the sun and hanging out with Sherman. Rather than boarding our gang miles away with the sheriff's deputy, Kip had arranged with his neighbor across the road to let us keep them in his meadow. Kip was right; posed against that great prairie skyline, with their winter shag gone and their muscles rippling beneath their gleaming coats, the donkeys really did look magnificent—although a little bewildered.

"Sherman has been glued to me," Zeke said with a laugh. "He's

Zeke and Sherman, reunited after the long trip from the Southern End

like, 'Thank God! One point of reference in this strange new universe that my brain can process!' " Zeke was processing a point of his own; despite his stubborn insistence back home in Lancaster that he still might be able to race, the throbbing in his leg after the past few days of travel was finally convincing him that it just wasn't possible. He was handling his disappointment like a champ, though; I was touched by the way Zeke had transitioned himself from Sherman's teammate into his support crew. Personally, I'd have been absolutely unbearable if I were in his boot, grumbling and moping around in everybody's way, but Zeke seemed genuinely jazzed about his new role as the only guy who could help his replacement get used to Sherman and make sure Sherman understood what was going on. Zeke wasn't just a new man; he was the new Tanya.

"So how does this work?" Kip asked. Even though they're locals, Kip and Kristin had spent so much time working overseas that they'd never actually been in Fairplay for a burro race. I was happy to let Zeke answer while I sank back and relaxed, ready for a little more shut-eye. Zeke filled them in on tactics and training, then Mika explained how Sherman had joined our family in the first place. Suddenly it got very quiet, as if everyone was holding their breath. I cracked an eye to see what was going on, and found Kristin and Kip swiveling their heads from Sherman to Mika, listening intently as Mika described Sherman's transformation from a sick, lame loner into this affectionate doofus who—fingers crossed—was about to compete in a World Championship mountain race.

"Wow," Kip finally sighed. "That is so punk rock." Kip had been a hardcore headbanger growing up, and he'd never forgotten the time a front man opening for Fugazi kept singing through the blood after someone in the mosh pit splattered his nose with a glass ashtray. "Idiot who threw that, that ain't punk," Fugazi's lead singer raged when he came onstage. "The guy you hit, he's punk. Being there for his band. Never giving up. Taking the pain. That's punk."

You're damn right, I thought as I lay back down in the sun. What's more punk than burro racing? Every other sport in America is about throwing the ashtray. They teach you to hit hard, be aggressive, play better than girls, never give up the ball. Power and possession, strength and domination: that's American sports in a nutshell. Then along comes this scruffy crew who give the finger to all that. Burro racing was inspired by prospectors and jackasses, the true American misfits, and it flips everything about modern sports on its head. Try any of that *No pain, no gain* stuff with a donkey, and you're in for a world of disappointment. You've got one hope of getting to the finish line, and that's to forget about dominance and ego and discover the power of sharing and caring, compassion and cooperation. That's not to say burro racers are powder puffs;

they're as fit as panthers and ferociously competitive, but the men still salute the women who beat them, and the women will run their guts out but still respect the Sisterhood of the Traveling Pee Break: "You never want to leave a competitor alone out there on the mountain," Barb Dolan explains. "So if one of the girls has to pee, we all stop and wait. After that, it's gangbusters." Taking the pain. Being there for your band.

Kip saw it immediately. Burro racing is a rebel yell from our outcast ancestors, a reminder that things used to be different—and it's not too late to go back.

"So who's going to fill in for Zeke?" Kip asked.

That flicked my eyes open. The day after Zeke broke his foot, I started calling everyone I could think of in Colorado in search of a sub. Turns out, we weren't the only burro racers who were jinxed that summer. Just about everyone I spoke to was dealing with some kind of injury, calamity, or Grade A ass-ache. Lynzi Doke had hurt her hip during track season and needed surgery. Barb Dolan's knees were bothering her, plus she'd been shocked when a dear friend suddenly dropped dead from an aneurysm. Barb decided it was time to step back and reassess, so she was re-re-retiring. Hal Walter, meanwhile, was at his wits' end with Teddy, the new half-wild burro he thought was going to be a speedster but was turning out to be a bit of a head case who freaked around water.

But I still had Wann card left to play: Brad Wann, the grizzly-tough dad who got into burro racing to help his epileptic son. Brad is a man-mountain of radiant energy who loves finding stone walls in his path so he can leave them with Brad-shaped holes. He's made it a point to become best buds with everyone in burro racing, if not most of the Southwest, so I was sure he'd tear into this challenge and come up with the perfect candidate to step in for Zeke. I thought it was pretty odd when Brad never responded to my first two messages, and just when I was considering a third, I heard from his wife, Amber. A week earlier, she told me, Brad had been overcome by a mystery illness that landed him in intensive care. His doctors had no idea what was causing the raging fever and fluids pooling in his lungs, and were keeping him alive with oxygen and a twenty-four-hour antibiotic drip. "It's breaking my heart to see my big guy wasting away and not 100 percent there in his thoughts," Amber told me. Their son, Ben, was also being hit by seizures again after a four-year respite. Between her two guys, Amber was living a nightmare of helpless uncertainty. Luckily, after Brad had lost a quarter of his body weight and undergone a battery of organ and toxicology tests, he began to recover and was finally able to come home. Ben's seizures were still worrying Amber sick, but at least her man was out of the woods.

I couldn't bring myself to even mention my problems to Amber. I was at the end of the road—until I realized I was looking down the wrong road. Just about every burro racer alive lives in Colorado, but there is one glorious exception: the clan Pedretti. Every year, as they have for more than a decade, Pedrettis of all ages caravan down from Wisconsin to compete in the World Championship in honor of the late Rob Pedretti. Out of all those nieces and brothers and cousins, there had to be at least one Pedretti they could spare, right? I got Rob's brother on the phone, and all I had to do was tell him the basics about Zeke and Sherman before he cut me off.

"Same problem Rob faced," Roger said. "Straight A's in school, amazing athlete, best mountain lion guide in the state, maybe the country . . ." Roger's voice trailed off. Suddenly, he snapped back. "And your burro! We can't let them down. No way." Roger thought for a moment. "You know what? My sister-in-law will run with you. Count on it."

"Roger, you're my hero," I said. "But don't we have to check with her? Just to make sure?"

"Nah," Roger said. "That will get my brother involved. Leave it to me."

. . .

"So that's where we are," I told Kip. "We're waiting for the Pedrettis to arrive so we can meet our Mystery Teammate and find out if she's really up for this. Nail-biting time."

"If you want, I could practice with you a little," Kip volunteered. He was in solid shape from mountain biking, and even though he hadn't run in a while, he could handle a few miles. I jumped at the offer. I wanted to see how Sherman would react to a new partner, and Mika and I had to at least try to get used to the high altitude. Four days isn't nearly enough time to adjust to 10,000 feet, but I was hoping for a little reassurance that our summer of hill running would soften the impact.

We hustled back to change into running gear, then gathered our ropes and harnesses and called the Gang of Three in from the meadow. The donkeys ambled over to the gate and held their heads patiently while we tacked them up. Already, Kip had the makings of a natural. During his time in Alaska, he'd done a fair bit of skijoring—cross-country skiing while pulled by sled dogs—so he'd developed an ease and quiet touch around animals. Once Zeke showed him how to rub Sherman's ears just right, it looked like Kip and the Wild Thing would get along fine.

"*Heey—*" I began, but Flower sprang into action before I finished, trotting down the long dirt road that stretched from the Earthship across the mesa to the foothills. Matilda must have been just as eager for action after the long trip. She and Mika locked in right beside me, while Sherman—

Vanished. In his place, a stone statue of a donkey appeared at the end of the Earthship's driveway. Kip did everything he could to urge him forward, but Sherman remained in character and didn't move a muscle. Zeke hobbled over on his crutches to help. He tried to lead Sherman for a few steps to get him started, and maybe it was my imagination, but I'd swear Sherman gave Zeke a look that said, "For real? You think I'm falling for that?" Sherman and Zeke were finally back together again. If we'd thought Sherman was just going to walk off with some guy he'd never met before, we didn't know Sherman.

Burro racing is so punk rock! Kip leads us on a pre-race run near his Earthship.

"Let's swap out and see if he goes with you," I told Mika. She took Sherman, and I pinned Flower right up beside him, with Kip and Matilda bringing up the rear. We started again, this time at a slow walk, allowing all three donks time to stay together. After a while, I eased Flower into an easy, shambling jog. Matilda followed right away. Sherman dropped his big Eeyore head and harrumphed, then fell in behind Matilda. After two days trapped in a truck, it was bliss to feel my body warm from the exercise and the summer sun—for about three minutes. I lifted my hand in surrender, too winded to say stop. My head was swimming so badly I was on the verge of hitting the dirt.

"Thank. God," Mika panted. "Dying."

Ken waited patiently while Mika and I, hands on our knees, fought to catch our breath. I remembered hearing that Paige Alms, the champion big-wave surfer, trained at this same altitude because it was akin to getting hammered to the bottom of the ocean by forty-foot waves. Even for a pro athlete in peak condition, it's excruciating. "Running at 10,000 feet is the hardest thing I've ever done," said Paige, and that was even after she'd survived enough bad falls to be a multiple nominee for Worst Wipeout of the Year. *Your sport*, Paige was reminding us, *is my punishment*.

Or death sentence. Not far from where we were, a twenty-year-old Pennsylvania woman was hiking near Aspen when she was overcome by altitude sickness. It sounds benign as a bellyache, unless you know that altitude sickness is actually shorthand for "High Altitude Pulmonary and Cerebral Edema." When the young woman reached 10,000 feet, her lips would have turned blue and she'd have coughed up bloody froth. She'd be wheezing for air and blinded by an excruciating headache. She would have to lie down—and that would seal her fate. Her lungs would fill with fluid, drowning her on dry land. At that moment, we were approaching the same elevation she was at when she died.

We started again, even more gently this time, but I managed only fifty yards before my pounding headache made me stop. Behind me, Mika was already walking, her head down and hands on her hips. Did I really want to put her through this—for a donkey race? And we still had to climb another 2,000 feet. What were we doing? And of course, the one person on our team who'd really dedicated himself to cross-adapting to high altitude by squatting in icy water was the one person who was sitting back at the house with his foot in a boot.

"This is looking like a bad idea," I told Kip. "I'll be honest, I'm afraid of what can happen on the mountain. We can really get in trouble up there."

"You're right," Kip said. "But you might want to think about this." He hesitated, weighing whether to go on. "Look, I hate to sound all mystic and shaman-ey," he continued. "There's this Burmese saying that always stuck with me: 'Fear that thing, do that thing.'"

"Yeah," I said, faking as much interest as I could muster. "Cool." Was he kidding me with that shit? I needed Advil and a blood transfusion, not some cornball fortune cookie. My irritation was reaching the boiling point—and then, just like pulling a sink plug, it drained away. Not because of anything I did, but because of Flower. After six months of burro-whispering, Flower had conditioned me to blow out tension as soon as it reached critical mass. *All right*, I told myself. *Calm your ass down.* I remembered something Curtis had told me: "Your mind will beat you before the mountain does."

Mika caught up with us. She flopped across Sherman's back, using him as a pillow while she took a breather. "What do you think?" I asked her. "Call it a day?"

"I'm going to be slow, but I can keep going," Mika said. "Maybe it will get better."

"That's the spirit," Kip said with a smile. "Can't get worse, right?"

I sucked in a big bellyful of air, hoping it would help, and it reminded me of Curtis's other advice: "Find the rhythm." Donkeys may not look like ballroom dancers, but if you watch closely, everything they do has a tempo. They trot and breathe to the beat of a waltz, and that's what allows them to keep going, lightly and easily, for miles and miles across sun-scorched canyons.

So this time I tried to sync myself to Flower. I focused my mind on her rhythm, and as we ran, I counted out a beat.

One, two, three . . .

One, two, three . . .

We made it 50 yards, then 50 more, before I realized the numbers had turned into words:

Fear that thing . . .

Do that thing . . .

Fear that thing . . .

Do that thing . . .

Two days later, a minivan pulled up to the Earthship and out piled a pack of Pedrettis.

"You definitely called the right Pedretti," Rick said. "My brother can talk my wife into anything. She'll come home with something she bought and say 'What do you expect? I was with Roger.'"

We liked Tammy immediately. She was joyful and friendly, and despite her delicate glasses and apologies in advance for holding us back—"Really, I'm *slow*"—it was apparent as soon as she slipped out of her Patagonia jacket that if we had one weak link in this operation, it wasn't Tammy. She looked more fit than any of us, except maybe Zeke, with the kind of ropey muscles in her shoulders you get when you'd rather be strong than skinny. Tammy had taken up running only a few years ago, signing up for her first half marathon as a fortieth birthday present to herself, and like us, she felt in over her head. But Tammy had witnessed a lot of Pedretti burro-racing horror stories over the years and was quick to reassure us that sooner or later, one way or another, everyone finds the way back. The first time her husband ran the long course, he spent an hour lying flat on his back on the rocky mountainside, eyes closed in exhaustion, until dive-bombing mosquitoes forced him back to his feet. The following year, Rick was stranded in freezing wind for an hour again because his burro refused to walk through snow. His brother Roger once finished with a broken rib after being dragged by Dakota, which wasn't so bad considering Rob had to be medevaced to Denver after he ran the back half of Leadville despite being kicked so hard his lung collapsed.

"You know how they say, 'It never always gets worse?'" Rick said. "It's a lie."

"Don't listen to him," Tammy said. "If I can do it, so can you." That hushed Rick, but more important, it opened our eyes to the fact that Tammy had actually run a World Championship before. *Twice*, it turned out. I didn't remember Roger ever mentioning that, and it gave us an instant double shot of hope. When it came to mountain training, Wisconsin was no better than Pennsylvania, right?

Sherman and the Gang were eyeing us curiously from the fence, so Mika and Zeke led Tammy and her kids over to meet them, while Rick and I walked back to the Earthship to gather the tack. "Can I ask what happened with Rob?" I said, once we were out of earshot. I'd heard bits of the story, but ever since Roger had made that comment about Zeke, I'd wanted to know more.

"It's okay, I like talking about him," Rick said. "Our brother was one of a kind."

Rob was a real intellect, he said, a super-quick learner with boundless energy. As a boy, he showed such a knack for animals that both grandfathers recruited him to help train their coonhounds. While other kids were asleep in bed, young Rob was plunging through the woods, following the baying of his dogs as they pursued raccoons in the dark. Rob ran cross-country in college, and became so hooked on endurance sports that after graduation, he moved to Colorado to join the growing tribe of roving ultrarunners. Right away, Rob got himself a pair of dogs, and even though nursing was his profession, he was soon in demand as a hunting guide. "Toughest little shit we ever met," one of Rob's hunting companions would say. "He was ridiculously strong for a small man. He'd throw a 150-pound mountain lion over his shoulders and still leave you behind. He was a natural for chasing lions, because those hunts can turn into marathons. One guy said, 'The best dog I have is Rob.' "

Rob's combination of brains, tenacity, and animal empathy was the perfect raw material for burro racing, and once he gave it a try, he was hooked. For the next five years, he and Samaritan were locked in mortal combat with Hal Walter and Tom Sobal in races that were often decided by no more than a donkey's nose. Hal and Rob were ferocious rivals but became devoted friends, and it was Hal who first detected something a little strange about Rob's behavior. "Rob was always on the go," Hal noticed. "Any downtime was filled with caring activities, such as coaching a Little League baseball team or working odd shifts as a nurse at nursing homes. Rob rarely spent much time at home," Hal realized. "I often wondered why."

Still, Hal was shocked when he heard that Rob, out of the blue, had suddenly sold his hunting dogs, closed his outfitting business, and was moving to St. Louis to become a chiropractor. He was thirty-six years old, Rob explained, so he'd decided it was time to quit running around like a kid and settle into a steady profession. "Suddenly, he's out of the mountains and in the city," his buddy

Kenny would tell me. "Only goal he's chasing was education, and that was no challenge at all." Rob blitzed through the classwork with ease, which left him with time on his hands and without any wilderness to explore or dogs to roughhouse. Tough little shits don't whine, though, so even as Rob began sinking into depression, he shared his despair only with his journal.

"He wrote about his pain in the last months, and when you read it you think, 'Man, how did he make it this far?' " Rick said. It was Rick who heard the gunshot that February day in 2004 and came out of the woods carrying his dead brother in his arms. "You never know what kind of demons somebody is facing," Rob's friend Kenny would later lament. "Because if Rob Pedretti isn't invincible, nobody is."

Rob was finally at peace, but his mother wasn't. Carol hated the idea of Rob's beloved racing burro waiting for him in a Colorado pasture, never knowing if Rob was coming back. Roger set off to fetch Samaritan with no truck, no trailer, and no clue how to operate either, but that turned out to be the least of his problems, because the guy who had Samaritan refused to give him up. It took three full years of relentless Pedretti-ing before Samaritan was finally handed over. Roger tried to step into Rob's shoes, and when he and Samaritan first ran the World Championship in Rob's honor, the family descended on Fairplay in such force that more Pedrettis showed up for the race than racers.

Privately, they're also here to settle a debt. Rob's mom spent three years fighting for rights to a donkey, and that's only a taste of what she'll do to show her gratitude to the community that helped her son when he was alone and far from home, struggling in silence against a disease that was trying to kill him. Rob didn't know he'd accidentally discovered a treatment for his depression by boosting his life-saving levels of serotonin and oxytocin. He just knew he'd suddenly been swept up by a burro-racing band of brothers and sisters who gave him a chance to run hard, breathe the snowy air, and care for the kind of big, furry creatures that have made humans feel safe and content since the dawn of time.

"Your friend Zeke," Rick concluded, hurrying to finish his story as we approached the gang in the pasture with the donkeys' gear. "This might be the best thing for him."

We could see Zeke across the field, surrounded by kids and up to his wrists in Sherman's ears as he showed everyone how to give Sherman his favorite deep-tissue massage. Sherman was adoring all the attention. He looked so happy these days—so strong and self-assured—that it was sometimes easy to forget that for all the healing he'd gotten from Zeke, he'd given as much in return.

"Anyone seen the Ladies?" I asked after we'd finished our warm-up session with Tammy and the Pedrettis had headed back to the Hand Hotel.

A chorus of nopes. No one had seen them since breakfast that morning—or, wait, wasn't that yesterday? They seemed to have vanished in the night, or at least before dawn, and now it was getting dark again. I'd been so distracted by our trial run with Tammy, I hadn't thought about the Ladies all day. We started Tammy off with Matilda, and right away she showed her chops. Tammy positioned herself right in the sweet spot near Matilda's flank, and urged her along with a steady stream of chirps and lip smacks. Matilda always likes working with a fellow boss, so she clicked with Tammy as quickly as we did. Even Sherman was surprisingly chipper. Maybe all that ear caressing in the meadow had won him over to Team Pedretti, because suddenly he was okay leaving Zeke for a while and trotting down the dirt road with Mika and a pack of Pedretti kids. We managed a gentle three miles, and it wasn't too bad; breathing was still torture, but whoever was trying to smother us to death with a pillow had loosened his grip.

"You're going to have a lot of fun," Rick promised, then clawed it back. "Well, not at the start. That's crazy. There's this one dirt bank, *really* steep, and suddenly you've got sixty burros crashing down that thing, stuff flying off their saddles . . ." Luckily, the younger Pedrettis were hungry and pulled Dad toward the car, nipping off any more blood-and-guts details about what we were facing. We made plans with Tammy to meet on race-day morning, then waved good-bye and headed toward the Earthship for a potluck dinner with Kip and Kristin's family.

I checked my phone: still no word from the Ladies. Part of me said chill; the Ladies had been gallivanting nonstop since we'd arrived in Colorado, and so far they'd never had a problem. Twice, they'd set off into the hills on horses borrowed from randos; once, they drove halfway across the state and back to have lunch with Linda's daughter, who lived a few hours from the Wyoming border. Always, they barreled in with tales—weird pawprints they'd seen, or hidden mountain passes they'd four-wheeled, or the free tattoo parlor they'd discovered in Fairplay (the catch: you do it yourself)—before offering to take the kids out back and teach them how to shoot cans with Karin's pistol.

But the more adventures they had, the more I wondered how long their luck could hold. They were so smart and tough that my imagination couldn't conjure up an image of them actually getting hurt. But landing inside a cell after telling a Colorado state trooper exactly where and how far up he could shove his ticket book? Snapping both axles in a 4x4 duel to the top of Mount Aspen? Picking up a teenage hitchhiker running away from an abusive home and masterminding her escape to a safe house in Portland? The only shocker about any one of those scenarios would be that it hadn't happened already. And where would that leave the rest of us? Stuck in the middle of a lonely mesa on Planet Earthship, with no way of getting the donkeys back home to Pennsylvania, let alone into Fairplay for the race. As much as I got a kick out of the Ladies' squeeze-the-day approach to life, I didn't want them tipping the canoe when we were *thisclose* to shore.

Finally, at eight p.m., I couldn't wait any longer, so I tried Karin's number. No answer. No voice mail. I clicked over to e-mail, since Karin knows I don't text. Nothing.

We'd already finished dinner and the kids were dealing out Uno cards when headlights finally appeared in the driveway. We uncovered the food, expecting the Ladies to be ravenous when they burst through the door, but they never appeared. I headed out and found them sitting silently in our adobe cabin, sharing a drink from a bottle of Bird Dog Whiskey.

"You're not hungry?" I asked.

"No, not tonight, sweetie," Linda said, in a faraway voice I'd heard from her only once before: that night in a West Virginia gas station when she told me how close they'd come to losing Karin.

"We're just going to sit here quiet for a bit," Karin said. "You can go on back. We're okay."

Any one of those things in itself was a red flag—sitting, quiet, or by themselves—but all three at once meant they were far from okay. The reason Karin hadn't answered the phone, she said, was because she was busy with the police. "We were driving back here, and we got passed by this Mustang, really flying," she said. Seconds later, a motorcycle screamed around them, hot on the Mustang's tail. The motorcycle rider tried to whip back into Karin's lane, but he cut too hard and his footrest caught the road. It jerked the bike back up, throwing the rider. He flew through the air, the bike tumbling around him, and Linda's heart froze: "No helmet. He was up in the air and I was thinking, 'Oh no, baby, don't hit the ground.'"

Karin stomped her brakes, smoking to a stop and skidding the truck sideways to block both lanes of the road. She threw open her door, then stopped to grab her phone while Linda—yes, seventy-two-year-old, eleven-year-old-girl-size Linda—ran into the face of oncoming traffic to cover the injured man with her body. He'd already stopped breathing, but when Linda knelt beside him, he gasped. Was his wife there? he asked. His wife was in the Mustang. He was trying to catch up with his wife. Was his wife there?

"She's coming," Linda said. "She'll be here."

The rider fell quiet. "Baby, keep breathing," Linda pleaded, her hands on his chest. "Baby, keep breathing." The Mustang's driver must have seen the crash in her rearview mirror, because moments later, the motorcyclist's wife ran to his side. She arrived as he took his last breath.

"By the time the police came, he was gone," Karin said.

She saw the look on my face and poured me out a sip of whiskey. "You could have been killed yourselves," I said. "You're lucky you weren't run down out there." For all the Ladies' bravado about grabbing their purses anytime they got wind of an adventure, I couldn't forget that the real reason Karin and Linda were there was because they were the only ones who stepped up when they heard we were in a jam. They'd never met us, but they signed on anyway for a grueling trip and a week living in a trailer. I knew they liked taking their own chances, but I hated knowing that because of me, they could have been dead on the highway thousands of miles from home.

"You've been great to all of us," I said. "I'm really sorry it led to this."

Karin stood up and put down her glass. "I learned something from being sick," she said, and began shaking her arms like she was shedding water. "When bad stuff is behind you, you've got to shake it off and move ahead. We've been through a lot this trip. Before we go home, let's make sure it's worth it."

26

An Army of Wann

Eight forty-five in the morning on July 31, 2016, will go down as the single greatest moment in Sherman's life. That's when he looked outside and discovered that donkeys had taken over the world.

We'd just turned right at the only traffic light in Fairplay and entered a town filled with jackasses. It was like the summoning of a barbarian army, all of them furry, sturdy, and intense, braying a battle cry that passed through the ranks and echoed through the streets. I glanced in the side-view mirror and saw that back in the trailer, Sherman was electrified: he'd spun around from his usual ass-backwards position toward the wall and shoved his head out the window, nostrils flaring and ears sky-high. He puffed his lips, like a tenor loosening up, then erupted in a yodel so thundering and soulful, it seemed to tell his entire life story. His tribe was here. Gathered to greet him. At last.

"This is probably as close as we can get," I told Karin. Mika and I had ridden into Fairplay with the Ladies, while all of our crew, including Zeke's and Kip's families, was caravanning behind. With little more than an hour left before the starting gun, livestock trailers had crowded every inch of the fairground and choked the side streets. "Once we get into town, we'll never have room to turn around."

"We're good," Karin said. "We've got VIP parking."

"We do?" Outside, I could see we were passing Curtis Imrie's trailer, unmistakable with its Old Glory red-white-and-blue paint job and emblazoned with his motto: "Donkeys, Drama & Democracy." "It doesn't look like Curtis has VIP parking," I pointed out, "and he's been doing this for forty years."

"Don't distract her," Linda scolded. Karin threaded the giant trailer through the needle of narrow streets, heading straight for the race registration table in front of the historic Hand Hotel. When we were about ten steps away from the starting line, she cut the wheel and eased the truck to a stop in front of a giant hand-painted sign reading: NO RACE-DAY PARKING. YOU WILL BE TOWED!!

"I don't think we're getting away with this," I said.

"Nah, we talked with the lady and she said we could," Karin said.

"What lady?"

"Such a sweetheart," Linda said. "We met her in the Italian restaurant and she said, 'Be my guest!'"

Sweet mystery of life. By that point, the only thing that surprised me about the Ladies was the way they kept finding ways to surprise me. If they lucked into the very best parking spot in a jam-packed town because they had an afternoon craving for calzone, well, of course they did. "We've gotta get a move on," Karin said, ending the discussion. "The more time the donkeys have to get used to this scene, the better." We hopped out of the truck and pulled open the trailer doors, watching as Sherman, Flower, and Matilda slowly stepped out, awestruck, and joined the Grand Gathering of Burro Nation.

Sherman was so excited, his fur seemed to be standing on end. Probably at no time since the Civil War have more donkeys been united in one place than here, at Colorado's annual burro races, and it was turning into a family reunion for creatures who never

On World Championship morning, Fairplay belongs to the burros.

knew they had cousins. Up and down the street, burros were hollering and sniffing and checking one another out. Sherman threw back his head and thundered out another yodel of joy, with Flower and Matilda joining in to lend backup vocals. Karin covered her ears. "Oh my god. It's like heavy metal out here."

Mika and the Ladies set out buckets of feed and fresh water for our Gang of Three, while I went to meet Tammy. I found the Pedretti clan, many still in pajamas and red Wisconsin sweatshirts, clustered in the Hand Hotel's breakfast room. Roger and Rick were the only ones not eating. They were in running shorts and warm-up tops, having just jogged down to check their burros at the fairground.

"They seemed edgy," Roger said.

"They know it's race day," Rick agreed.

"Worst two hours of the year," Roger said, holding a hand over his belly as if he were sick.

"Suddenly, the air is a whole lot thinner."

Right—the air. Last night, I'd woken up over and over, thinking about the girl from Pennsylvania who'd gone for a hike near here and died at 10,000 feet. One of the scary things about altitude sickness is the way it fogs your mind before it kills you; just when red flags are going up that you're in trouble, your brain is too oxygen-starved to recognize them. I knew that Carol, the Pedretti family matriarch, was a nurse, and by now she'd seen many rookies from the flatlands attempt this race.

"Should I be worried?" I asked her. "Is there a way to know if things are getting serious before your lips turn blue?"

"I've got just the thing," Carol said. She rummaged in her bag, pulled out a finger-clamp blood pressure monitor, and reached for my hand. Her friend Renee is also a nurse, and she leaned over with Carol to see the readout. "People all over the country run a marathon and go 'Woo hoo!,' but nothing compares to this," Renee said. "These people are *tough*. So if you're not up to it," she added, laying a consoling hand on my arm, "you shouldn't feel bad."

"Ninety-one over seventy-seven," Carol said. "Oh, that's good. You're good."

"And this really works?" I asked. The Pedretti boys looked at each other and shrugged. "We're still here," Roger said.

Tammy appeared as I was plucking my finger out of the gadget. If anyone on your team feels disoriented, Carol warned, stop immediately. Don't lie down; lean against your burro and stay on your feet until help comes. "You've been here a week, so you should be fine," she concluded. "But you can't take any chances. If it hits you, it's for real." With that send-off ringing in our ears, Tammy and I headed back to the trailer to gather our saddles for weigh-in and registration.

Every racer's saddle had to be equipped with the traditional trio of prospector's tools—pick, pan, and shovel—while burros of Flower's size had to carry a weight minimum of thirty-three pounds. Zeke's parents had arrived the night before, just in time for Zeke's dad, Andy, and Kip to tear into our equipment in Kip's bike shed. I'd heard nightmare stories of racers who'd fought their way through snowdrifts and hailstorms for twenty-nine miles, only to approach the finish line and have to turn around when they realized they'd shed a few tools back on the trail. Andy and Kip made sure our equipment was bolted firmly to the saddles, then they tightly wired a pair of fifteen-pound barbells to Flower's pick and shovel to reach the required thirty-three pounds. Our rigs were bombproof.

As Mika, Tammy, and I were shouldering our ropes and saddles to have them weighed and measured, Hal Walter suddenly appeared. He came stalking toward us, dressed in black running tights and a black jacket, his head hanging and his face bristling with stubble, looking as distant and distracted as when we'd first met in Leadville ten years earlier.

"Hey, Hal," I called. His head jerked up, and he managed a thin smile. "Good to see you," he said.

"You okay?" I asked.

"Bad week," he grimaced. "I've been sick. Really worn out and stuffed up. And then this weird vein bulged out on my chest. I noticed it in the shower yesterday. I tried pushing it back in, but it popped out again. I don't know what the hell it is. Spent all last night researching it online. By the way," he added, suddenly shifting gears, "do you know if they're measuring ropes?" I had to guess that besides his sudden-onset vein tumescence, Hal was also worried about Teddy, his feisty new burro that was so rambunctious around water that he'd taken a chunk out of Hal's shoulder while crossing a creek. "He lashed out like a snake and went for blood," Hal said. "Bit me so hard, blood began gushing. Kind of traumatic, really. I thought he'd severed some tendons." He sighed. "How long have I been doing this? Thirty-seven years? How come it never gets easier?" Because of Teddy's issues, Hal was wondering if he could get away with a longer-than-rulebook fifteen-foot rope to reduce his risk of getting dragged.

"I don't know," I said. "We haven't checked in yet."

"Yeah, never mind. I'll see you up there. Or not. I don't know."

At the registration desk, I hung my saddle on the scale and bent down to sign the waiver. For choice of race, we'd decided on the fifteen-mile course. It was farther than we'd ever gone in training, and the idea of attempting to double that distance after the summer we'd endured seemed not only crazy but pointless. We'd had one goal from the beginning: to give Sherman a job he could love and friends he could share it with. If we could push ourselves halfway up the side of that mountain and bring him back, still trotting strong, that would be a win.

"Did Hal say where he's going?" the race director asked. "We're about to start and he's not signed in."

"I don't think he's going to run," I said. "He's having some health issues."

"'Course he is," the race director said. "We call him Nervous Nellie. Hal has won this race more than anybody, and every year he's a wreck. Watch—he'll forget his health issues and be right in the fight."

Sure enough, Hal came slouching to the table before I was finished with the paperwork. Pushing him from behind was Curtis Imrie. "Look who I found," Curtis said. "Better brand him and get him in with the herd before he breaks loose again." When Curtis saw me, he left Hal and wrapped me in a big hug. "You did it!" he said. "You jumped all the way in and committed to some burros. Be the death of you."

"Are you racing today?" I asked, noticing the big hinged brace on Curtis's knee.

"Sure. You work with what you got."

The countdown clock was ticking. We hurried back to the trailer, and with the Ladies' help, we saddled the donkeys and double-checked their halters. I pulled on a pair of running sandals I'd been given by my *Born to Run* friend, Barefoot Ted. I didn't know if they were ideal for this terrain, but it felt smart to arm ourselves with charms and talismans. Mika wore her Speedy Goat Farm tee as a reminder of home, and Sherman had the special brown rope with a tasseled edge that we'd given Zeke for his birthday.

Sophie and my niece, Sara, clustered around Sherman and gave him a good deep head scratching for luck. "I hope you like your birthday present, Soph," I said. It suddenly seemed a lifetime ago since we'd first seen Tanya crashing out of the woods on her riding donkey, Muffin, and Sophie had thought she might like one herself. "If it wasn't for you, Sherman would still be locked in that stall." Suddenly, my stomach knotted so badly, I could barely breathe. Hal Walter had won this thing seven times and it still scared him. Was Sherman really ready for this? Were any of us?

"This is it," I croaked. "Everyone set?"

"Hang on," Zeke said. He stepped up and rubbed Sherman's ears. "Some last lovin' from Papa Zeke."

"He wouldn't be here without you," Mika said, pulling Zeke in for a hug. "None of us would."

Zeke's eyes misted. He blinked fast and looked away. "Are you thanking me, or blaming me?" he said, as his mom came over to throw an arm around his shoulders.

"We'd better get moving," I told Mika and Tammy. We took the donkeys by the halters and marched them toward the crowd gathering at the monument to Prunes, Fairplay's beloved town burro. Prunes had wandered loose in the 1860s after working in the mines, but instead of disappearing into the mountains, he stuck around town, circling the streets with such a routine that people on one side of Fairplay could stick notes on his halter for friends on the other. Prunes died in 1930, but he had become such a part of Fairplay's identity that his memorial has been preserved in a place of honor to this day.

"CHRIS!" someone was shouting. "We made it!" All around us, burros and racers were shifting and circling like an agitated sea. I scanned around and finally, at the far edge of the crowd, I spotted Amber Wann jumping up and down. Beside her was her husband, Brad, hollow-cheeked after his hospital nightmare but still beaming, happy to be back among his fellow misfits. Amber pointed, and I saw their thirteen-year-old son, Ben, nervously holding a burro's lead rope.

"Watch out for him!" Amber called, crossing her fingers. For the first time in his life, Ben was attempting the race alone. Amber had decided they wouldn't run that year because of Brad's illness, but Ben was adamant: epilepsy or not, he was going to get out there and show what Wanns are made of. Amber asked Curtis's opinion, although she should have known better. "Someone has to be young and foolish. Might as well be Ben. Can't always be me," Curtis said. Then he added quietly, "Mrs. Wann, you know that every man and woman on that mountain will watch over Benjamin like their own child. Starting with me."

We tried to move back toward Ben, but the donkeys were getting antsy and twisting around one another, tangling ropes and clanging packsaddles. Rick forced his way through the crowd for one last bit of advice. "This place is about to erupt—" he began.

"RACERS! ARE YOU READY?" the race director shouted.

Rick raised his voice. "Whatever you do—"

The mob began to chant: "TEN . . . NINE . . ."

"Hold them back!"

"SIX . . . FIVE . . ."

"If you go anaerobic in this race—"

"THREE . . . TWO . . ."

"Your race is over!"

BLAM!

The dam burst, and donkeys flooded Front Street. The lead pack blasted off at a gallop, with elite runners like Justin Mock and George Zack relying on honed marathoner's speed to keep pace, while the not-so-elites who'd gotten sucked in beside them were hauling back on their ropes and wondering how much longer they could hold on. Beside us, a woman was at war with a bucking bronco; her burro reared and kicked as she circled at a distance, gripping her rope heroically while trying to soothe it back down. I searched quickly for Ben Wann, but he'd disappeared in the mayhem. Even if I'd spotted him, I had my hands full with Flower.

Flower was quivering with anticipation, dying to join the fun, powerful enough to break free at any moment. I spun her backward so she was facing away from the stampede, a potentially winning strategy if she hadn't kept spinning until she was pointed forward again. So I met her halfway and hoped for the best; gripping her by the halter, I let her tow me along at a fast walk, trying to hold her back from a gallop. I glanced behind and saw a donkey I didn't recognize; the Wild Thing was politely declining an opportunity to create more havoc than he'd ever dreamed possible and was instead strolling calmly beside Matilda, the two of them taking in the mayhem like a pair of spectators enjoying a piece of performance art.

Clattering hooves approached hard from behind, and then Rick and Roger pulled alongside us with their big racing burros. Flower tugged to run beside them, so I took a chance; I let my hand slip down Flower's halter and played out the rope, testing whether I could manage their pace. And as soon as I did, I was the one who needed to be held back; it was a total rush to shift into a run and join the herd rather than resist it. Instead of being battered by monster waves, it was like riding one, sensing its power to beat you down but knowing you'll be whisked softly to shore. Side by side with the Pedretti boys, we climbed to the top of Front Street and then, in a weird footnote to an already weird event, we entered South Park City, a restored mining town said to be an inspiration for Comedy Central's *South Park*.

"Control," Rick reminded me. He pointed to a runner clipping along just ahead with a mini-donkey no bigger than Matilda. "You can learn from that guy, John Vincent."

"Is he the guy Barb Dolan calls 'that fucking leprechaun'?"

"Barb calls him 'that fucking' lots of things," Rick agreed. "He aggravates her. But he's good."

That was Rick's Wisconsin-nice way of saying "See ya!" He and Roger floated ahead, shifting up to gears we couldn't match. Fast as the Pedrettis were, though, John Vincent and little Crazy Horse continued to pull away even as the brothers accelerated. Still, John couldn't shake a young woman who trailed barely two steps behind, matching him stride for stride. I'd heard about that phenomenon: she had to be Louise Kuehster. Like Lynzi Doke, Louise had taken up burro racing in high school, and now that she was a twenty-year-old freshman rower at the University of Oklahoma, she'd returned to the mountains with formidable skills, strength, and speed.

Even from a distance, Louise and John's battle was beautiful to behold: their legs swung in perfect unison with the burros', their feet *pop-pop-pop*ping lightly off the ground like boxers skipping rope. Their flow reminded me of Curtis's tip about running to the rhythm, and just in time: at half a mile into the race, the altitude was catching up with me. I sucked in a steadying breath and calmed my pace to the beat of Kip's mantra: *Fear that thing . . . do that thing . . . Fear that thing . . .*

"Here she goes again," I yelled back to Mika and Tammy. "Everyone good?"

"So far!" Mika said.

The best thing about Flower, as the race was teaching us, also made her a pain in the ass. Flower has a social instinct that's off the charts; she's gentle with everyone and obsessive about sticking tight to Sherman and Matilda. She's also a stalker who beelines obsessively for anyone in front of her. That instinct was terrific for training; back home, we could always send one of the kids out on her bike as a rabbit, and Flower would follow so closely that her breath would puff the rider's hair. But out here on the mountain, it was rabbits as far as the eye could see. By this point, we'd run about a mile, and the field of racers had strung out into a long parade heading up into the mountain. For Flower, every runner ahead of us was even more irresistible than the one we'd just passed.

But I had to admit, I loved her guts. Flower was pushing me to try something I would never have done on my own: compete. Two days ago, Mika and I were panting our guts out in Kip's driveway, wondering how we would ever be able to run more than a

With Tammy Pedretti taking Matilda, Team Sherman is off and racing.

few dozen yards at 10,000 feet. Our strategy for today was to trot slowly and be happy with finishing dead last, as long as we finished. Something was clicking, though; every time Flower set her sights on another burro farther up the trail, I would prepare for dizzy spells and wobbly legs. But so far, nothing. Either we were jacked through the roof on adrenaline and a contact serotonin high from all the animals around us, or we'd acclimated just enough for the benefits of Coach Eric's hillwork-hell month to kick in.

We were rocking along so well, I was napping when Flower led me straight toward death. "Dude!" I blurted. "What the hell!" The race course had exited the woods and, for a short stretch, skirted the highway. Before I noticed what she was doing, Flower curved us off the trail and right toward the stream of speeding cars. I pulled her back and started off again, but three yards later, she veered toward the road again like she was drawn by a suicidal impulse. Was her stalker reflex short-circuiting, somehow attracting her to the noise and energy of whooshing cars? I had no idea, but luckily, we had a remedy for mysteries like this.

"Time for Matilda," I called.

Tammy jogged up at once with her partner. "Can you see if she'll lead us out of this?" I asked. "I don't know what's wrong with Flower." Tammy gave a little Barb Dolan growl, and Matilda lunged obediently ahead, swerving around Flower to take point.

Sherman bolted to catch up, leaving Flower to traipse along behind as Tammy led us out of the danger zone. Soon we were back in the brush and I could risk letting Flower set pace as lead dog again. The trail was twisty and tough to follow, with deep ruts and scraggly rocks hiding everywhere as broken-ankle traps. I was so intent on keeping my eyes down and my focus sharp that I was shocked when we rounded a bend and suddenly I was face-to-face with Kip.

"Five miles!" he hooted.

"Seriously? We've gone five?" I looked around. "Where's everyone else?"

"That boy, Ben, was having trouble with his burro, so the girls were back helping him out."

Ten seconds earlier, we'd been hucking our way through the brush and hating life. Now that we knew how far we'd come, I didn't want to stop. "Wait for the kids, or keep going?" I asked Tammy and Mika.

"Keep going!" they both sang out.

I couldn't get over the all-terrain skill show that Mika and Sherman were putting on behind me. Sherman was so enthralled to be flowing along in a donkey parade, he was nearly bouncing—and that still made him only the second-happiest being attached to his rope. Every time I glanced back, Mika was chatting to Sherman, urging him onward with a steady stream of endearments. I didn't know if Mika was riding an adrenaline high or was finally getting the payoff she deserved from all our hill training in the form of a red blood cell boost, but she was having the run of a lifetime. Suddenly, it dawned on me how fitting it was for Mika to be sharing this moment with Sherman.

We'd all been so focused on Zeke's accident that until then it never clicked that Mika had undergone a transformation just as big as Sherman's, and all for only one reason: to help everyone else. Me, Zeke, Sherman—we'd all gotten into this thing because there

Mika trains with Flower and follows the skirt-and-a-smile strategy.

was something about ourselves that needed fixing. Mika had been doing just fine. She'd never asked for any of this, but all it took for her was one look at a sick donkey with crippled feet, standing alone in the back of a hay wagon, and she was committed to doing whatever would make him well again. During all those months of bucking hay in 20-degree temperatures, and hauling buckets of water from the creek in snowshoes, and waiting for me to figure out how the hell to get us back out of the maze, she was never anything less than joyful. No one had worked harder this past year, or come farther, than the two of them. Zeke was Sherman's best friend, no doubt about it; but Mika was the loving spirit that made this possible for all of us.

Half a mile later, I heard the low rumble of hoofbeats and quickly pulled back on Flower's rope. The lead racers were flying back toward us, and I was thrilled to see Louise hadn't lost a step on past champion John Vincent. The two of them were pinballing expertly around the serpentine trail, pitter-pattering over the rocks and ruts without a stumble. With little over five miles to go, Louise looked strong enough to blow past him at any moment.

"GO, LOUISE!" I shouted, and got a quizzical smile in return. *Thanks, buddy. And who are you again?*

After the front runners had whooshed by, we chirruped the donkeys back into a trot and soon popped out of the woods and onto a dirt road dropping downhill to the turnaround. I felt something bumping and snorting against my leg, and looked down to find Sherman—Sherman!—bursting ahead to challenge Flower for the lead. Until then, he'd been cruising steadily, but seeing the lead donkeys streaming back toward us put a zip in his stride and made his ears rocket up, as if he was inspired by his first good look at what a fast donkey could do. We rolled on down the hill, arriving at the halfway point and its authentic burro race–style aid station: a battered cooler with a few water bottles and a trough for the donkeys.

All six of us took a few quick sips, then five of us headed for home. I didn't realize what was going on until I'd gotten a good hundred yards or so back up the homeward climb and heard the bellow of a heartbroken Sherman echoing behind me. I turned back and saw that Sherman hadn't budged from the cooler. Mika was trying everything to get him moving again, and when she saw we were looking, she raised her hands in a big shrug: *What now?* Tammy and I turned around—but then I heard Karin's voice in my mind. "See it through his eyes," she would say. So what was Sherman up to? He wanted Flower and Matilda to meander around with him for a while at the trough. Going back wouldn't change that. Only one thing would.

"Let's go," I told Tammy. "We're leaving them."

She cocked an eye. "You're leaving your wife? On the mountain?" Not even the speediest women in the Sisterhood of the Traveling Pee Break would abandon one of their own out here, and I was ditching my spouse? "She'll understand," I said. And hoped. I straightened Flower around and hopped her up again. Tammy and I got back in our groove, and when Sherman hollered this time, I kept eyes front. Sherman let loose another heartbreaker; it lingered and echoed, and then everything went silent. I cranked my head as little as possible, just enough to sneak a glance back, and saw the Wild Thing hammering toward us at a gallop. Mika had played out every inch of her rope and was barely holding on to the end.

"Shit, now we've got to go back," I told Tammy, wheeling Flower around. If Mika and Sherman broke contact, she wouldn't be able to continue the race until she'd caught him herself and returned to the spot where he'd broken free. With more than seven miles yet to run, we couldn't risk draining our tanks by chasing Sherman around in the Rockies. Tammy and I hurried back downhill, and when Sherman saw us coming, he eased to a trot. Mika managed to stay with him until we were all united again. We gave Sherman a chance to nip and frisk around with his buddies while Mika fought for her breath, and then we started again for town.

Ahead, we could hear yelling and a piercing whistle. Ben Wann was struggling with Burrito, I saw, and out of nowhere a band of spectators had magically materialized and were cheering him along. As we got closer, I realized it wasn't magic so much as Kip's driving: he'd manhandled our rented minivan up that rutted fire road and flung the doors open, allowing Kristin and Zeke and the kids to pour out and let Ben know that no matter where he was out here, friends were near. We crested the hill just as Ben was heading down, sharing weak high-fives on both sides. We now had less than five miles to go, but we'd also crested 11,000 feet and it was catching up with us. We were about to leave the fire road, and I was dreading what lay ahead.

"I can't take another stretch in the maze," I muttered to myself, before realizing what I was thinking. *The maze?* Did I really just blank out and think I was back in the Southern End, fighting our way through another steaming summer workout in the slate quarry? I forced my mind back to my conversation with Carol that morning, trying to remember what she'd said about altitude sickness. Were delusions a red alert? I put up a hand, calling for a slow-down, before I realized there was another side to this. I was exhausted; I was sucking in big belly breaths; but I was also imprinted by the memory of that first day in the maze, when we'd led the donkeys into the woods with no idea how we'd get back out again. I'd never forgotten that moment, the do-or-die sensation that if we were serious about tackling the World Championship, the maze was a fear we were going to have to conquer. No wonder it came roaring back up again when I was feeling like these last few miles might be more than we could handle.

"Okay, I'm good now," I said, and turned toward Mika. "Ready for some fun in the maze?"

"Always," she said, and that was the last word out of any of us for the next few miles. We dropped our heads and locked eyes on the trail, willing our minds to stay sharp as we felt our energy fading. It felt like forever before we looked up—and saw a horrifically steep hill ahead. It looked like torture. And then I remembered. "Is that the dirt bank that Rick told us about?" I asked Tammy.

"I think so. Yes, definitely. That's it."

Then that means . . .

Together, we clawed and scrambled our way up the slippery bank, our feet churning through crushed dirt as soft as sand. The donkeys lunged and powered up on their own, the tools clanging against their saddles. I slipped to my knees, and when I saw that Mika was about to lose her footing, I jammed my hand under her sneaker as a foothold. I'd never felt more desperate to get up a hill, because I had a feeling that at the top we'd find . . .

Rick was right. That hill was a beast, but when we beat it, we were looking down at the last quarter mile along Front Street to the finish line. "Ready for your victory parade, Sherman?" I said. Sherman was shaking his mane and nipping at Flower, impatient for these slow humans to get going. After all, he had a job to do. "Let's do it, Flower," I called, then pulled her back. If anyone was truly reaching a finish line today, it was Mika and Sherman. I wasn't sure how Sherman would handle the commotion of a cheering crowd, but as Mika took him out and we trotted down the street, his head flashed up. I shouldn't have worried about letting Mika go first: as soon as Sherman saw Zeke waiting at the finish line, he was off like a shot . . .

Except Matilda wouldn't have it. She coiled and sprang, whipping past Sherman and leading us across the line in four hours and two seconds, putting us in 28th, 29th, and 30th place out of fifty-two starters. I dropped my hands to my knees, exhausted and elated, until I looked up and saw Amber and Brad Wann. We'd been through a lot with Sherman that year, and there were times when it felt like we were struggling with a challenge that we would never really understand. But that was nothing compared to what they must be going through. The Wanns had put their hopes on a burro, and now their boy was out there by himself while they waited, anxious and helpless, hoping they were right.

As I was struggling for something to say, Amber's eyes lit up. Far up Front Street, a small figure was jogging steadily closer. Murmurs spread through the crowd, and then a cheer grew until it turned into a roar. By the time Ben crossed the finish line, the noise was too deafening to hear the announcer shout his name. But everyone could read the words on Ben's shirt, and in a flash, I realized what Sherman must have been thinking from the moment he got back on his feet and was given a chance to start his new life:

Trust Me. I've Got This.

tracks, across a few creeks, and into the back pastures of at least half a dozen farms. For the first time since well before her accident, Tanya would be reunited with Flower, while I'd see what I could do with Chili.

"You're okay riding?" I asked, as Tanya pulled herself onto Flower's back.

"As long as it's on this sweetie," she said. "Donkeys are so much steadier than horses, so I should be fine."

We were silent for the first mile or so as I wobbled along behind Tanya, doing my best to remember her tips and match her poise, screwing my butt hard into the saddle and keeping my heels down. I could barely relax long enough to take a breath from one end of the maze to the other, but after I survived that roller coaster, I felt comfortable enough to pull up alongside Tanya and chat. Unbelievably, she had suffered yet another horrendous burst of bad luck: during a thunderstorm, a utility pole had crashed down in her corral and electrocuted her favorite carriage horse. Tanya nearly died herself when she raced outside at the sound of the transformer bursting and remembered only at the last second not to touch the metal gate. But recently, there were some glimmers that her insane cycle of misfortunes was coming to an end. She was recovering well from her crash, partly because her riding tutorials with the teenage neighbor were strengthening her back and reviving her spirits by getting her out in the woods. She was still scrambling to support her farm, but she'd become a fixture in the local Amish community and had a full slate of driving clients to keep her afloat.

She loved hearing about her rough-riding soul sisters, the Ladies, who'd bonded with us so tightly as adventure buddies that I was planning to see them in a few weeks if I learned to trail-ride well enough to join them for an event in Virginia. Tanya was eager for updates on her old pal Zekipedia, who was now back at Penn State with his one-eyed cat studying time travel, or whatever nano/neuro/nuclear stuff he majored in. I was delighted to report that Zeke had strayed from his marriage with physics to date an actual human woman, although of course the information was delivered Zeke-style: "She's a great mathematician," he'd told me, before revealing under follow-up questioning that oh, yeah, she's also lovely and warm and witty. Zeke's sister, Ashling, was also in a great place and had done so well at Penn State that she was selected for a fellowship in pharmacology at one of Philadelphia's most prestigious hospitals.

I told Tanya all about those badass young burro-racing women. Louise Kuehster was edged out by John Vincent in the World Championship, but came back two weeks later to smoke him in Buena Vista. Meanwhile, Lynzi Doke—who should have been dead twice—had put her future as the next Barb Dolan on hold for the time being while she focused on school and track. She would graduate as one of her school's best athletes and at the top of her class, and go on to follow her mother into nursing.

But that winter, Lynzi lost one of her biggest fans: Curtis Imrie, the beloved chieftain of the burro racing tribe, died suddenly of a heart attack while leading one of his prize burros at the National Western Stock Show in Denver. Hal Walter didn't know how he'd break the news to his son, Harrison. So many times, it was sharp-eyed Uncle Curtis who first noticed that Harrison was on the verge of an episode, and he'd suddenly shout, "You startin' something, boy? Let's take this outside!" The battered old cowboy and the struggling eleven-year-old would charge through the door, tumbling down into the dirt to rassle around furiously until, eventually, whatever was boiling inside Harrison's mind had eased and disappeared. Together, Hal and Harrison worked through their heartache. Running helped, and by the time Harrison reached high school, he'd become a phenomenon. "Harrison is now running varsity track," Hal would proudly report. "He runs the 400, the 800, the 1600, and the 3200 all in one meet. No other kids do this." Harrison puts on a big pair of noise-canceling headphones so he isn't triggered by spectator sounds, and since he can't hear the starter's gun, he watches the front foot of the runner beside him. Off the track, his teammates are fiercely protective. "There's a senior named Kyleigh Martin who raises steers, really quite the

Hal and Harrison during a track meet

cowgirl," Hal told me. "No one would ever bully Harrison for fear she'd stomp their ass." Hal himself had become so adept at guiding Harrison through the stresses of a track meet, he was hired as head coach, and soon other neurodiverse students were following Harrison's lead and joining the team. Harrison can still be a volcano—recently, a track teammate had to talk him out from under a desk—but Hal can't get over the way his life has changed. "Community, goals, friendships, a social life," Hal marvels. "Who knew that running would be his ticket to belonging to something?"

Between the stories and the gorgeous trail ride, Tanya and I were having a great morning—

Until I told her about the day that Sherman's owner came looking for him.

I was working in the back field one afternoon when I got a glimpse between the trees of someone who'd appeared behind the house and was up to something at the fence. I took my time going down, figuring it was one of the local kids who'd stopped by to see the goats, but as I got closer, I suddenly felt a stab of apprehension: the hoarder was here. He and his wife and daughter were leaning over the fence, snapping their fingers and calling out to Sherman. I quickly counted back over the months; was it two years already? No, not for a while—but for someone who was obsessed, that would be a technicality.

"Shaggy!" they called. "Don't you want to come say hello?"

Sherman stood his ground, tucked between Flower and Matilda some fifty feet from the fence. The hoarder looked around when he heard me coming through the gate. There was no way in hell that Sherman was leaving our care, so I knew this wasn't going to be pretty. The hoarder's wife spoke up first and told me that because her husband loved animals so much, they'd taken him to a small zoo in Maryland for his birthday. On the way back, they realized they were passing the home of a church member who'd negotiated Sherman's liberation and remembered I lived somewhere just down the road. They drove over and spotted Sherman. As they were talking, they kept looking back at Sherman and trying again to entice him to the fence. Sherman just stared, never moving a muscle.

"He looks so good," the hoarder's wife said. "Doesn't he look good, Dad?"

"I wish he'd come over and say hi," the hoarder replied. The look on his face was so sad, so genuinely bewildered and bereft, that any blame I'd laid on him for Sherman's condition instantly disappeared. He was so infatuated with animals and so delighted to be around them, he couldn't see that his affection had become a disease. He and his family could clearly tell how much better off Sherman was in his new home, though, and they were happy for him. They hadn't come to take him back. They'd come to say good-bye.

"Lucky I wasn't there," muttered Tanya, who's a lot slower than I am to forgive even well-intentioned sins against the animal kingdom. We finished the last long uphill climb to her house, then slid stiffly out of the saddle, both of us sore to the bone and regretting

Back home and waving to a neighbor during a run with Sherman

that we hadn't shut up a while ago and brought our ridiculously long ride & chat to an end. Neither one of us really had the energy to wrestle Flower into the trailer, but amazingly, she ambled right in as soon as Tanya opened the door.

"Well, well," Tanya said, impressed. "You must have taught her a thing or two."

But my secret was blown a few minutes later, when we came over the hill toward home and Flower saw Sherman and Matilda waiting for her at the fence. Flower brayed in excitement, and then they all cut loose, the three of them yodeling a donkey song of love. Tanya smiled and shot me a glance. The only thing I'd really taught Sherman, she knew, was that he'd never be alone again.

Running with Sherman

Acknowledgments

Just as I was pulling into my dentist's parking lot one afternoon I got a call from Tara Parker-Pope, my editor at *The New York Times.* Tara was inviting me to speak to her journalism students at Princeton, and before we got off the phone, she asked, "What are you up to these days?" She expected to hear about writing, but what was on my mind was this sick donkey we were struggling to keep alive. "That's going to be a great book," Tara said, and when I told her no, there's nothing to write about because I didn't even know if it was going to survive, she replied, "*That's* why it's a great story." She went on to suggest I write a weekly column for her about animal-human partnerships, and that's how "Running with Sherman" began. It's also where I got the title: I kept peppering Tara with alternative names, and she just shushed me and said, "Trust your editor. Call it 'Running with Sherman.'"

I still wasn't sure how a book would work out until I spoke with Richard Pine, my agent at Inkwell Management. Richard is a nice guy, but he ain't too nice to pull his punches. He doesn't mind telling me when I'm wrong—I'm pretty sure he actually enjoys it—and he not only perked up about the thought of a donkey book but also came up with the subtitle: "The Donkey with the Heart of a Hero."

Perfect.

We floated the idea by Edward Kastenmeier, my editor at Knopf who has now guided me through three projects over the course of thirteen (!) years, and, as usual, he saw the possibilities immediately. This might be the strangest endeavor we've tackled together because the story travels in so many directions and involves so many people close to my heart, and I'll always be grateful for the brilliant guidance Edward gave me along the way. He was also clever enough to hire Caitlin Landuyt, the point person at Knopf who had to handle all my last-minute changes and corrections and photo additions. Whatever Knopf is doing to attract such cool, intelligent people, it's working.

Somehow, I lucked into the same professional romance on the other side of the Atlantic. I've been with Profile Books since before anyone had ever heard of *Born to Run*, and I couldn't dream of a better partnership. Andrew Franklin and his team treat me like family, which means they're warm and encouraging but also ruthless about working me to the bone when it's time to hit the pavement and sell some books. I'll miss Profile master publicist Anna-Marie Fitzgerald while she's off having a baby (hooray!), but I'm sure by the time you read this, they'll already have found another human cyclone to step in.

I've tried to make sure that everyone who rallied to help Sherman is reflected in these pages, but behind the scenes there are still many quiet heroes who deserve special applause. Like Don Korenkiewicz and Ruby Rublesky and Steve Farrah, who came to the rescue when Zeke and I were busy breaking various body parts and we needed emergency substitutes to help us train the donkeys. Our neighbors have made the Southern End a magical place to live for nearly twenty years and never blinked when they suddenly began spotting us running down the road every morning with the Gang of Three. The Boomsma family and the Metzler clan are always there for us when we need help, and somehow we always do. Gini Woy not only took the magnificent cover photo of Sherman, but her daughter, Stella, should seriously consider dropping out of high school to pursue a career in professional burro racing. She'd be a superstar.

I've told you a lot in here about Curtis Imrie, but that's only a slice of the appreciation he deserves. When we lost him, it was like a portion of the Earth had fallen away. To Curtis and all the members of the burro racing community, Sherman and I can never thank you enough.

ILLUSTRATION CREDITS

The photographs in this book are used by permission and courtesy of the following:

Amber Canterbury: p. 203
Courtesy of Amos King: p. 136
Andrea Cook: pp. 297, 314
Chancey Bush/Evergreen Newspapers: p. 240
Christopher McDougall: pp. 7, 12, 40, 117, 166, 265, 274, 282, 292, 293, 324
Dana McDougall: p. 322
Dave Epper: p. 53
Ed Kosmicki/photosourcewest.com: p. 236
Hal Walter: p. 193
Jennifer Russ: p. 332
Julie Angel: pp. 33, 111
Kelly Doke: p. 196
Luis Escobar: p. 73
Matt Roth: pp. 31, 85, 334, 336
Mika McDougall: pp. 101, 145, 294, 302

ALSO BY

CHRISTOPHER MCDOUGALL

NATURAL BORN HEROES

Mastering the Lost Secrets of Strength and Endurance

Christopher McDougall's journey begins with a story of remarkable athletic prowess: on the treacherous mountains of Crete, a motley band of World War II Resistance fighters abducted a German commander from the heart of the Axis Occupation. To understand how, McDougall retraces their steps across the island and discovers ancient techniques for endurance, sustenance, and natural movement that have been preserved in unique communities around the world. His search takes us scrambling over rooftops with a Parkour crew in London, foraging for greens with a ballerina in Brooklyn, tossing heavy pieces of driftwood on a Brazilian beach with the creator of MovNat—and, finally, to our own backyards. *Natural Born Heroes* will inspire readers to unleash the extraordinary potential of the human body and climb, swim, skip, throw, and jump their way to heroic feats.

Sports/Adventure

ALSO AVAILABLE

Born to Run